Introduction

The *Primary Mathematics* curriculum allows students to develop their ability in mathematical problem solving. This includes using and applying mathematics in practical, real-life situations as well as within the discipline of mathematics itself. Therefore the curriculum covers a wide range of situations from routine problems to problems in unfamiliar contexts to open-ended investigations that make use of relevant mathematical concepts.

An important feature of learning mathematics with this curriculum is the use of a concrete introduction to the concept, followed by a pictorial representation, followed by the abstract symbols. The textbook does supply some concrete introductory situations, but you, as the teacher, should supply a more concrete introduction when applicable. The textbook then supplies the pictorial and abstract aspects of this progression. For some students a concrete illustration is more important than for other students.

This guide includes the following :

- ◆ **Scheme of Work**: A table with a *suggested* weekly schedule, the primary objective for each lesson, and corresponding pages from the textbook, workbook, and guide.

- ◆ **Materials**: A comprehensive list of manipulatives and other material used in this guide, and a list of material at the beginning of each chapter.

- ◆ **Objectives**: A list of objectives for each chapter.

- ◆ **Vocabulary**: A list of new mathematical terms for each chapter.

- ◆ **Notes**: An explanation of what students learned in earlier levels, the concepts that will be covered in the chapter, and how these concepts fit in with the program as a whole.

- ◆ **Activity**: Teaching activities to introduce a concept concretely or to follow up on a concept in order to clarify or extend it.

- ◆ **Discussion**: The opening pages of the chapter and tasks in the textbook that should be discussed with the student. A scripted lesson is not provided. No guide can anticipate every student's responses. The material is meant to give you sufficient background to discuss the concepts based on your student's needs.

- ◆ **Practice**: Tasks in the textbooks students can do as guided practice or as an assessment to see if they understand the concepts covered in the teaching activity or the discussions.

- ◆ **Workbook**: Workbook exercises that can be done after the lesson.

- ◆ **Reinforcement**: Additional activities that can be used if your student needs more practice or reinforcement of the concepts. This includes references to the exercises in the optional *Primary Mathematics Extra Practice* book. Use at your discretion for whether you think they will be useful for your student.

- ◆ **Games**: Optional simple games that can be used to practice skills.

- ◆ **Enrichment**: Optional activities that can be used to further explore the concepts or to provide some extra challenge.

- ◆ **Tests:** References to the appropriate tests in the optional *Primary Mathematics Tests* book. See below for a description of the test book.

- ◆ **Answers:** Answers to all the textbook tasks and workbook problems, and many fully worked solutions. Answers to textbook tasks are provided within the lesson. Answers to workbook exercises for the chapter are located at the end of the chapters in the guide.

- ◆ **Mental Math**: Problems for more practice with mental math strategies.

- ◆ **Appendix**: Pages containing drawings and charts that can be copied and used in the lessons.

The textbook and workbook both contain a review for every unit. You can use these in any way beneficial to your student. For students who need more continuous review, you can assign three problems or so a day from one of the practices or reviews. Or, you can use the reviews to assess any misunderstanding before administering a test. The reviews, particularly in the textbook, do sometimes carry the concepts a little farther. They are cumulative, and so allow you to refresh your student's memory or understanding on a topic that was covered earlier in the year or in a previous year.

In the *Primary Mathematics Tests* book, there are two tests for each chapter and two cumulative tests for each unit. The second test of each set is multiple choice. You can use the material in this book for tests, review, or extra practice. You can combine tests and do them at the end of the unit rather than after each chapter. There are plenty of choices for assessment, review, and practice.

The mental math exercises that go along with a particular chapter or lesson are listed as reinforcement or enrichment in the lesson. They can be used in a variety of ways. You do not need to use all the mental math exercises listed for a lesson on the day of the lesson. You can have your student do one mental math exercise a day or every few days, repeating some of them, at the start of the lesson or as part of the independent work. You can do them orally, or have your student fill in the blanks, depending on the type of problems. You can have your student do a 1-minute "sprint" at the start of each lesson using one mental math exercise for several days to see if he or she can get more of the problems done each successive day. You can use the mental math exercises as a guide for creating additional "drill" exercises.

The "Scheme of Work" on the next few pages is a *suggested* weekly schedule to help you keep on track for finishing the textbook in about 18 weeks. No one schedule or guide can meet the needs of all students equally. For some chapters, your student may be able to do the work more quickly, and you can combine lessons, particularly those that are review of earlier material. For other lessons, take the time your student needs on each topic and each lesson. For students with a good mathematical background, each lesson in this guide will probably take a day, along with a test some days at the beginning of the next lesson, depending on how many of the optional material you use.

The 2011 printing of this guide was written when the 2008 and 2009 printings of the textbook, workbook, *Extra Practice*, and *Tests*. New printings of these books may have corrected errors or may have slight changes.

Scheme of Work

Textbook: *Primary Mathematics Textbook* 5B, Standards Edition
Workbook: *Primary Mathematics Workbook* 5B, Standards Edition
Guide: *Primary Mathematics Home Instructor's Guide* 5B, Standards Edition (this book)
Extra Practice: *Primary Mathematics Extra Practice* 5, Standards Edition
Tests: *Primary Mathematics Tests* 5B, Standards Edition

Week		Objectives	Text book	Work book	Guide
Unit 6 - Decimals					
		Chapter 1: Tenths, Hundredths and Thousandths			1-2
1	1	♦ Review decimals. ♦ Relate each digit in a decimal to its place value.	8-10	5	3-4
	2	♦ Compare and order decimals. ♦ Write decimals as fractions.	11-12	6	5
		Extra Practice, Unit 7, Exercise 1, pp. 141-146			
		Tests, Unit 7, 1A and 1B, pp. 1-7			
		Chapter 2: Approximation			6
	1	♦ Round a decimal to the nearest whole number. ♦ Round a decimal to 1 or 2 decimal places.	13-15	7	7
		Extra Practice, Unit 7, Exercise 2, pp. 147-148			
		Tests, Unit 7, 2A and 2B, pp. 9-12			
		Chapter 3: Add and Subtract Decimals			8
	1	♦ Add decimal numbers. ♦ Subtract decimal numbers. ♦ Check the reasonableness of the sum or difference.	16-17	8	9-10
		Extra Practice, Unit 7, Exercise 3, pp. 149-150			
		Tests, Unit 7, 3A and 3B, pp. 13-16			
		Answers to Workbook Exercises 1-4			11
		Chapter 4: Multiply and Divide Decimals by a 1-Digit Whole Number			12
2	1	♦ Multiply and divide a decimal number by a 1-digit whole number. ♦ Use estimation to check reasonableness of answers.	18-19	9	13
	2	♦ Round the quotient to 1 or 2 decimal places.	19-20	10-11	14

Week		Objectives	Text book	Work book	Guide
	3	◆ Express a fraction as a decimal correct to 2 decimal places.	20-21	12	15-16
	4	◆ Practice.	22		17
		Extra Practice, Unit 7, Exercise 4, pp. 151-152			
		Tests, Unit 7, 4A and 4B, pp. 17-20			
		Answers to Workbook Exercises 5-7			18
		Chapter 5: Multiplication by Tens, Hundreds or Thousands			19
3	1	◆ Multiply a decimal by tens.	23-25	13	20
	2	◆ Multiply a decimal by hundreds or thousands.	25-26	14-15	21
		Extra Practice, Unit 7, Exercise 5, pp. 153-154			
		Tests, Unit 7, 5A and 5B, pp. 21-24			
		Chapter 6: Division by Tens, Hundreds or Thousands			22
	1	◆ Divide a decimal by tens.	27-29	16	23-24
	2	◆ Divide a decimal by hundreds or thousands.	29-30	17-18	25
		Extra Practice, Unit 7, Exercise 6, pp. 155-156			
		Tests, Unit 7, 6A and 6B, pp. 25-28			
		Answers to Workbook Exercises 8-11			26
		Chapter 7: Multiplication by a 2-Digit Whole Number			27
4	1	◆ Multiply a decimal by a 2-digit whole number. ◆ Estimate the product in multiplication of decimals.	31-32	19	28-29
	2	◆ Multiply a decimal by a 2-digit whole number.	32	20-21	30
		Extra Practice, Unit 7, Exercise 7, pp. 157-160			
		Tests, Unit 7, 7A and 7B, pp. 29-31			
		Chapter 8: Division by a 2-Digit Whole Number			31
	1	◆ Divide a decimal by a 2-digit whole number. ◆ Estimate the quotient in division of decimals.	33-34	22-23	32-33
		Extra Practice, Unit 7, Exercise 8, pp. 161-162			
		Tests, Unit 7, 8A and 8B, pp. 33-38			
		Chapter 9: Multiplication by a Decimal			34
	1	Multiply a decimal by a "1-digit" decimal.	35-37	24-25	35-37
	2	Multiply a decimal by a "2-digit" decimal.	37	26	38
		Extra Practice, Unit 7, Exercise 9, pp. 163-164			
		Tests, Unit 7, 9A and 9B, pp. 39-42			

Week		Objectives	Text book	Work book	Guide
		Chapter 10 - Division by a Decimal			39
5	1	◆ Divide a decimal by a "1-digit" decimal.	38-39	27-28	40-41
	2	◆ Divide a decimal by a "2-digit" decimal.	40	29	42
	3	◆ Practice.	41		43
		Extra Practice, Unit 7, Exercise 10, pp. 165-166			
		Tests, Unit 7, 10A and 10B, pp. 43-46			
		Answers to Workbook Exercises 12-18			44
		Review 7	42-43	30-33	45
		Tests, Units 1-7, Cumulative Tests A and B, pp. 47-52			
		Answers to Workbook Review 7			46
Unit 8 - Measures and Volume					
		Chapter 1: Conversion of Measures			47-48
6	1	◆ Convert a decimal measurement to a smaller unit or a compound unit.	44-45	34	49-50
	2	◆ Convert a measurement to a larger unit as a decimal.	46	35-36	51-52
	3	◆ Practice.	47		53
		Extra Practice, Unit 8, Exercise 1, pp. 171-172			
		Tests, Unit 8, 1A and 1B, pp. 53-56			
		Chapter 2: Volume of Rectangular Prisms			54
	1	◆ Review volume.	48-52	37	55
	2	◆ Find the side of a rectangular prism given its volume and the other two dimensions or area of one face. ◆ Find the side of a cube given its volume.	52-53	38	56
7	3	◆ Solve problems involving rectangular prisms and the volume of liquids.	54-55	39-40	57
	4	◆ Solve word problems involving volume or rectangular prisms.	56-57	41-42	58-59
		Extra Practice, Unit 8, Exercise 2, pp. 173-182			
		Tests, Unit 8, 2A and 2B, pp. 57-65			
		Answers to Workbook Exercises 1-6			60
		Review 8	58-60	43-46	61-62
		Tests, Units 1-8, Cumulative Tests A and B, pp. 67-73			
		Answers to Workbook Review 8			63

Week		Objectives	Text book	Work book	Guide
Unit 9: Percentage					
		Chapter 1: Percent			64
	1	♦ Read and interpret a percentage of a whole. ♦ Express a fraction with a denominator of 10 or 100 as a percentage.	61-62	47-48	65
8	2	♦ Express a decimal as a fraction in simplest form.	63	49-51	66
		Extra Practice, Unit 9, Exercise 1, pp. 185-190			
		Tests, Unit 9, 1A and 1B, pp. 75-79			
		Chapter 2: Writing Fractions as Percentages			67-68
	1	♦ Express a fraction with a denominator less than 100 as a percentage.	64-65	52-53	69-70
	2	♦ Express a fraction with a denominator greater than 100 as a percentage.	66	54-55	71
	3	♦ Solve word problems that involve finding the percentage of the whole.	67	56-57	72
	4	♦ Practice.	68		73
		Extra Practice, Unit 9, Exercise 2, pp. 191-196			
		Tests, Unit 9, 2A and 2B, pp. 81-84			
		Answers to Workbook Exercises 1-5			74
		Chapter 3: Percentage of a Quantity			75-76
9	1	♦ Find the value for a percentage of a whole.	69-70	58-59	77-78
	2	♦ Solve word problems that involve percentage of a whole.	70-71, 73	60-61	79
	3	♦ Solve word problems that involve percentage tax, interest, discount, increase, decrease.	71-73	62-64	80
		Extra Practice, Unit 9, Exercise 3, pp. 197-200			
		Tests, Unit 9, 3A and 3B, pp. 85-88			
		Answers to Workbook Exercises 6-9			81
		Review 9	74-75	65-68	82
		Tests, Units 1-9, Cumulative Tests A and B, pp. 89-96			
		Answers to Workbook Review 9			83

Week		Objectives	Text book	Work book	Guide
Unit 10 : Angles					
		Chapter 1: Measuring Angles			84
10	1	♦ Estimate and measure angles in degrees.	76-77	69-72	85-86
	2	♦ Tell direction in relation to an 8-point compass. ♦ Determine the angle between various points on the compass.	77	73-74	87
		Extra Practice, Unit 10, Exercise 1, pp. 211-212			
		Tests, Unit 10, 1A and 1B, pp. 97-102			
		Chapter 2: Finding Unknown Angles			88
	1	♦ Recognize some angle properties involving intersecting lines. ♦ Find unknown angles using angle properties of intersecting lines.	78-81	75-76	89-90
		Extra Practice, Unit 10, Exercise 2, pp. 213-214			
		Tests, Unit 10, 2A and 2B, pp. 103-106			
		Chapter 3: Sum of Angles of a Triangle			91
	1	♦ Recognize some angle properties involving triangles.	82-84	77-78	92
	2	♦ Find unknown angles using angle properties of triangles.	85	79	93
		Extra Practice, Unit 10, Exercise 3, pp. 215-218			
		Tests, Unit 10, 3A and 3B, pp. 107-110			
		Chapter 4: Isosceles and Equilateral Triangles			94
11	1	♦ Recognize some angle properties involving isosceles and equilateral triangles.	86-88	80-81	95
	2	♦ Find unknown angles using angle properties of intersecting lines and triangles.	88-89	82-83	96
		Extra Practice, Unit 10, Exercise 4, pp. 219-222			
		Tests, Unit 10, 4A and 4B, pp. 111--116			
		Answers to Workbook Exercises 1-8			97
		Chapter 5: Drawing Triangles			98
	1	♦ Construct a triangle with given measurements.	90-92	84	99
		Extra Practice, Unit 10, Exercise 5, pp. 223-224			
		Tests, Unit 10, 5A, pp. 117-118			

Week		Objectives	Text book	Work book	Guide
		Chapter 6: Sum of Angles of a Quadrilateral			100
	1	◆ Recognize some angle properties involving quadrilaterals. ◆ Find unknown angles using angle properties of quadrilaterals.	93-94	85	101
		Extra Practice, Unit 10, Exercise 6, pp. 225-226			
		Tests, Unit 10, 6A and 6B, pp. 119-122			
		Chapter 7: Parallelograms, Rhombuses and Trapezoids			102
	1	◆ Recognize some angle properties involving parallelograms and trapezoids.	95-97	86-87	103
12	2	◆ Find unknown angles using angle properties of parallelograms, trapezoids, and triangles.	98	88-90	104
		Extra Practice, Unit 10, Exercise 7, pp. 227-230			
		Tests, Unit 10, 7A and 7B, pp. 123-128			
		Chapter 8: Drawing Parallelograms			105
	1	◆ Construct parallelograms with given measurements.	99-102	91-92	106
		Extra Practice, Unit 10, Exercise 8, pp. 231-232			
		Tests, Unit 10, 8A, pp. 129-130			
		Answers to Workbook Exercises 9-13			107
		Review 10	103-105	93-96	108
		Tests, Units 1-10, Cumulative Tests A and B, pp. 131-137			
		Answers to Workbook Review 10			109
Unit 11: Average and Rate					
		Chapter 1: Average			110-111
	1	◆ Find the average of a set of data. ◆ Find the average when given the total and the number of items.	106-108	97-100	112-113
13	2	◆ Find the total when given the average and the number of items. ◆ Solve problems that involve averages and measurement in compound units.	108-109	101-102	114
	3	◆ Solve word problems of up to 3-steps that involve averages.	109-110	103	115
		Extra Practice, Unit 11, Exercise 1, pp. 235-238			
		Tests, Unit 11, 1A and 1B, pp. 139-142			

Week		Objectives	Text book	Work book	Guide
		Answers to Workbook Exercises 1-3			116
		Chapter 2: Rate			117-118
	1	♦ Understand rate as one quantity per unit of another quantity.	111-112	104-105	119
	2	♦ Solve word problems that involve rate.	113	106-107	120-121
	3	♦ Solve word problems that involve rate.	114	108-109	122
14	4	♦ Solve rate problems that involve steps in rate.	115-116	110-111	123
	5	♦ Practice.	117		124
		Extra Practice, Unit 11, Exercise 8, pp. 239-242			
		Tests, Unit 11, 2A and 2B, pp. 143-146			
		Answers to Workbook Exercises 4-7			125
		Review 10	118-121	112-115	126-127
		Tests, Units 1-11, Cumulative Tests A and B, pp. 147-154			
		Answers to Workbook Review 11			128
Unit 12: Data Analysis					
		Chapter 1: Mean, Median and Mode			129
	1	♦ Find the mean, median, and mode of a set of data. ♦ Understand how mean, median, and mode differ in the information they provide.	122-123	116-117	130-131
		Extra Practice, Unit 12, Exercise 1, pp. 249-250			
		Tests, Unit 12, 1A and 1B, pp. 155-159			
		Chapter 2: Histograms			132
15	1	♦ Categorize data into intervals and display the data using a histogram. ♦ Analyze and interpret data from histograms.	124-127	118-121	133
		Extra Practice, Unit 12, Exercise 2, pp. 251-252			
		Tests, Unit 12, 2A and 2B, pp. 161-166			
		Chapter 3: Line Graphs			134
	1	♦ Represent data in a line graph. ♦ Analyze and interpret data in line graphs.	128-130	122	135-136
		Extra Practice, Unit 12, Exercise 3, pp. 253-254			
		Tests, Unit 12, 3A and 3B, pp. 167-174			
		Answers to Workbook Exercises 1-3			137

Week		Objectives	Text book	Work book	Guide
		Chapter 4: Pie Charts			138
	1	♦ Understand pie charts.	131-132	123-126	139
	2	♦ Interpret pie charts with fractions or degrees.	133	127-129	140
	3	♦ Interpret pie charts with percentages.	134	130-133	141
		Extra Practice, Unit 12, Exercise 4, pp. 255-260			
		Tests, Unit 12, 4A and 4B, pp. 175-179			
		Answers to Workbook Exercises 4-6			142
16		**Review 12**	135-139	134-138	143-144
		Tests, Units 1-12, Cumulative Tests A and B, pp. 181-187			
		Answers to Workbook Review 12			145
Unit 13: Algebra					
		Chapter 1: Algebraic Expressions			146-147
	1	♦ Use letters to represent unknown numbers. ♦ Write algebraic expressions. ♦ Evaluate algebraic expressions using substitution.	140-143	139-141	148-149
	2	♦ Use exponents in algebraic expressions. ♦ Evaluate algebraic expressions using substitution.	144-145	142-143	150-151
	3	♦ Simplify algebraic expressions in one unknown.	146-147	144-145	152
17	4	♦ Practice.	148		153-154
		Extra Practice, Unit 13, Exercise 1, pp. 269-274			
		Tests, Unit 13, 1A and 1B, pp. 189-192			
		Answers to Workbook Exercises 1-3			155
		Chapter 2: Integers			156
	1	♦ Review positive and negative integers. ♦ Represent positive and negative integers on a horizontal or vertical number line. ♦ Compare positive and negative integers.	149-151	146-147	157
	2	♦ Add positive or negative integers.	152	148	158
	3	♦ Add positive and negative integers. ♦ Relate adding a negative integer to subtracting a positive integer of the same numerical value.	153-154	149-150	159
	4	♦ Practice.	155		160
		Extra Practice, Unit 13, Exercise 2, pp. 275-278			
		Tests, Unit 13, 2A and 2B, pp. 193-196			
		Answers to Workbook Exercises 4-6			161

Week		Objectives	Text book	Work book	Guide
		Chapter 3: Coordinate Graphs			162
18	1	♦ Identify and graph ordered pairs in the four quadrants of the coordinate plane.	156-157	151	163
	2	♦ Graph simple linear equations using a table of values. ♦ Graph vertical and horizontal lines. ♦ Interpret information from the graph of a linear equation.	158-161	152	164-166
	3	♦ Practice.	162-163	153-154	167
		Extra Practice, Unit 13, Exercise 3, pp. 279-282			
		Tests, Unit 13, 3A and 3B, pp. 197-204			
		Answers to Workbook Exercises 7-9			168
		Review 13	164-169	155-160	169-170
		Tests, Units 1-13, Cumulative Tests A and B, pp. 205-212			
		Answers to Workbook Review 13			171
Answers to Mental Math					172-173
Appendix - Mental Math					a1-a6
Appendix					a7-a19

Materials

Whiteboard and Dry-Erase Markers
A whiteboard that can be held is useful in doing lessons while sitting at the table (or on the couch). Students can work problems given during the lessons on their own personal boards.

Round counters
To use as place-value discs, unless you buy commercial place-value discs, or make paper ones, or just draw them.

Place-value discs
Round discs with 0.001, 0.01, 0.1, 1, 10, or 100, written on them. You can label round counters using a permanent marker. You need about 30 of each kind.

Base-10 set
A set usually has 100 unit-cubes, 10 or more ten-rods, 10 hundred-flats, and 1 thousand-block. These are used rarely at this level, so are optional.

Centimeter cubes
The unit cubes from a base-10 set are very close to centimeter cubes.

Multilink cubes
These are cubes that can be linked together on all 6 sides.

Measurement tools
Ruler
Meter stick
Liter measuring cup
Protractor

Graph paper

Supplements
The textbook and workbook provide the essence of the math curriculum. Some students profit by additional practice or more review. Other students profit by more challenging problems. If the material in this guide along with the *Extra Practice* and *Tests* book are not enough, there are several supplementary workbooks available at www.singaporemath.com.

Unit 7 – Decimals

Chapter 1 – Tenths, Hundredths and Thousandths

Objectives

- Review decimals.
- Relate each digit in a decimal to its place value.
- Compare and order decimals.
- Write decimals as fractions.

Material

- Place-value discs for 10, 1, 0.1, 0.01, 0.001
- Base-10 blocks, optional

Vocabulary

- Decimal
- Decimal point
- Tenths place
- Hundredths place
- Thousandths place

Notes

In *Primary Mathematics* 4B students were introduced to decimal numbers to thousandths. This chapter is a brief review.

Place-value notation makes numbers understandable, and computation accurate and simple. We use ten digits to write numbers. The place of each digit in the number is significant; each digit has a value that is ten times as much as if it were in the place to the right of it and one tenth as much as if it were in the place to the left of it. The digits in the number 23,456 represents 2 ten thousands, 3 thousands, 4 hundreds, 5 tens, and 6 ones. The place value of the digit 3 is thousands, and its value is 3000. Each whole number can be expanded as the sum of multiples of the value for each place — 1, 10, 100, 1000, etc. So, 23,456 can be written as 20,000 + 3000 + 400 + 50 + 6.

Decimal numbers are an extension of place-value notation to include place values less than 1. We write a **decimal point** to the right of the ones place. The digits to the left of the decimal point are whole numbers; the digits to the right of the decimal point are decimals. A digit in the first place to the right of the decimal point, the **tenths** place, has a value that is one tenth of the value of the same digit in the ones place. A digit in the second place to the right of the decimal point, the **hundredths** place, has a value of one tenth of the same digit in the tenths place. A digit in the third place to the right of the decimal point, the **thousandths** place, has a value of one tenth of the same digit in the hundredths place. And so on.

1.234 is 1 + 0.2 + 0.03 + 0.004, or $1 + \dfrac{2}{10} + \dfrac{3}{100} + \dfrac{4}{1000}$. It can also be expressed as a mixed number, $1\dfrac{234}{1000}$. We can read this number as a fraction, *one and two hundred thirty-four thousandths*, or more simply, *one point two three four*.

If a decimal number is less than 1, we usually use a 0 as a place holder in the ones place. Writing 0.12 rather than .12 makes it easier to see, and pay attention to, the decimal point.

Usually, decimal places are not called by name after the thousandths place; we do not normally say "the hundred-thousandths place" but rather the "fifth decimal place." At this level, your student will mostly encounter decimal numbers to the third place, or thousandths.

Concrete manipulatives or pictorial representations, such as place-value discs, fraction squares, and number lines are useful to explain place-value concepts for decimal numbers. Since this chapter is a review, the pictures in the textbook should be sufficient. If not, add in actual place-value discs or use base-10 blocks where the thousands cube now represents 1. A unit cube is then a thousandth. You can ask your student to imagine the unit cube expanded to the size of the large thousand cube. If you think your student will have any trouble with this review, start with examples using base-10 blocks or place-value discs before looking at the examples in the text.

The example below shows 52.031 with place-value discs on a place-value chart.

Tens	Ones	Tenths	Hundredths	Thousandths
10 10 10 10 10	1 1		0.0 0.0 0.01	0.001
5	2	0	3	1

Decimal numbers are ordered in the same way as whole numbers. We start by comparing the digits in the highest place value. If they are the same, we then compare the digits in the next higher place value, and so on. As with whole numbers, we have to be careful to pay attention to the place value of the digits, not the number of digits. 4.5 is larger than 4.25 even though it has fewer digits.

The lesson in the textbook starts with an example using measurement. Decimals are primarily used in measurement so that the size, weight, or volume of small objects can be expressed in terms of standard units.

(1) Review decimals and place value

Discussion

Remind your student that fractions count parts of a whole. Decimals are fractions where each part is divided into 10 equal parts. The whole is divided into tenths, each tenth into tenths, which is the same as the whole divided into hundredths, and each hundredth into tenths, which is the same as the whole divide into thousandths. This allows us to write numbers smaller than 1 whole using the same decimal system we use for whole numbers.

$$\frac{1}{10} \text{ of } 10 = 1$$

$$\frac{1}{10} \text{ of } 1 = 0.1$$

$$\frac{1}{10} \text{ of } 0.1 = 0.01 = \frac{1}{100} \text{ of } 1$$

$$\frac{1}{10} \text{ of } 0.01 = 0.001 = \frac{1}{100} \text{ of } 0.1 = \frac{1}{1000} \text{ of } 1$$

Concept p. 8

The top of the page relates decimals to measurements. As you discuss this page, use "zero point eight three" for 0.83, but use "eight tenths" for 0.8 and "three hundredths" for 0.03 in order to emphasize the place value of the digits.

The whole is 1 m. The 8 in 0.83 stands for 8 out of 10 parts of the whole, or 8 tenths. The 3 in 0.83 stands for 3 out of 10 parts of one of the tenths, or 3 tenths of a tenth. If all the tenths were divided into 10 parts, there would be 100 parts. So the 3 in 0.83 also stands for 3 hundredths. Together, the 83 in 0.83 stand for 83 out of 100 equal parts. 8 tenths is the same as 80 hundredths. This is also illustrated with the fraction square at the bottom of the page. The square has been divided into hundredths. Each row or column is a tenth.

Tasks 1-3, pp. 9-10

These tasks emphasize that the base-10 nature of our numbering system extends to decimal numbers. Your student should know:

♦ We use only 9 non-zero digits.

♦ In order to write larger numbers, 10 ones are grouped into 1 ten and so a digit in the tens place stands for that many groups of 10 ones. 10 tens are grouped in to 1 hundred and a digit in the hundreds place stands for the number of groups of 100. And so on.

♦ Each digit is a tenth of the same digit one place to the left, and each digit is ten times the same digit one place to the right.

1. (a) 0.35
   (b) 3.7
   (c) 2.04

2. (a) 0.024
   (b) 0.315
   (c) 4.002

3. (a) 0.005
   (b) 2: 20; 4: 0.4; 3: 0.03
   Or 2: 20; 4: $\frac{4}{10}$; 3: $\frac{3}{100}$

♦ This is also true for numbers after the decimal point. As shown in Task 1, 1 one is the same as 10 tenths and 1 tenth is the same as 10 hundredths. As shown in Task 2, 1 hundredth is the same as 10 thousandths.

If your student has any difficulties with these tasks, which is unlikely, draw a place-value chart with columns for ones, tenths, hundredths, and thousandths on a personal whiteboard or paper and have her use actual place-value discs or draw them on the chart. That will make it easier to see, for example, why 2 ones 4 hundredths = 2.04 with a 0 in the tenths place; there are no tenths.

Make sure your student understands that adding 0's after the last non-zero decimal place does not change the value of the number. 4.1 = 4.10 = 4.100.

Tell your student that place values to the right of the decimal do not stop at thousandths. Write 0.0001 and ask her what place the digit 1 is in. It is in the ten-thousandths place. Ask her what the next place would be (the hundred-thousandths place). After a while it gets cumbersome to name the places by the fractions, so we just say the fourth decimal place and know it is $\frac{1}{10,000}$ of 1 (note the 4 0's in the denominator). There will be some problems in this unit with a digit in the fourth decimal place.

> $0.\underline{0001} = \frac{1}{10,000}$ of 1
>
> Ten thousandths place
> = fourth decimal place

Practice

Tasks 4-6, p. 10

These tasks will help you determine how well your student remembers decimal concepts. If he has trouble with any of them, you can have him use place-value discs on a place-value chart, and then have him do some additional similar problems without discs.

> 4. (a) 5.63
>   (b) 5.61
>   (c) 4.537
>   (d) 4.535
>
> 5. (a) 0.148
>   (b) 0.048
>   (c) 0.008
>
> 6. (a) 0.1
>   (b) 0.006
>   (c) 30

Workbook

Exercise 1, p. 5 (answers p. 11)

Reinforcement

Write some equations such as the following and have your student supply the decimal number.

- 0.44 + 4 + 40 + 0.004 = ?   44.444

- $6 + 0.2 + 0.007 + 30 + \frac{1}{100} = ?$ 36.217

- $100 + \frac{3}{10} + 0.02 + 8 = ?$ 108.32

- $\frac{34}{100} + 100 = ?$ 100.34

- 450 tenths = ?   45

- 450 hundredths = ?   4.5

- 450 thousandths = ?   0.45

- 3.333 = ? thousandths   3333

(2) Order decimals and convert to fractions

Discussion

Tasks 7-8, p. 11

Task 7: This task should not be difficult since each division in all three number lines is 0.001. Remind your student that numbers on a number line increase in size from left to right. 5.65 is greater than 5.63.

7. (a) A: 0.004 B: 0.007 C: 0.011 D: 0.013 E: 0.019
   (b) A: 8.004 B: 8.007 C: 8.011 D: 8.013 E: 8.019
   (c) A: 5.634 B: 5.637 C: 5.641 D: 5.643 E: 5.649

8. (a) 42.54
   (b) 63.182

Task 8: Remind your student that we compare decimals the same way as we compare whole numbers; by comparing the digits in the highest place values first. So 42.326 is smaller than 42.54, even though it has more digits, because the tenths digit is smaller in 42.326 than in 42.54. If necessary, you can have your student illustrate these numbers with place-value discs.

Practice

Tasks 9-11, pp. 11-12

9. (a) >   (b) <
   (c) <   (d) >
   (e) >   (f) <

10. (a) 3.02, 0.32, 0.302, 0.032
    (b) 2.628, 2.189, 2.139, 2.045

11. (a) 0.538, 0.83, 3.58, 5.8
    (b) 9.047, 9.067, 9.074, 9.076

Discussion

Tasks 12 and 14, p. 12

These tasks review converting a decimal to a fraction or mixed number in simplest form. Remind your student that fractions can be simplified in steps. In Task 12, Both 52 and 1000 are even numbers, so a logical first step is to divide both the numerator and denominator by 2. Then 26 and 500 are also even numbers. In Task 14, both the numerator and denominator are divisible by 5.

12. $\dfrac{13}{250}$ $\dfrac{52}{1000} = \dfrac{26}{500} = \dfrac{13}{250}$

14. $2\dfrac{9}{200}$

Practice

Tasks 13 and 15, p. 12

Workbook

Exercise 2, p. 6 (answers p. 11)

13. (a) $\dfrac{1}{2}$ (b) $\dfrac{2}{25}$

    (c) $\dfrac{1}{4}$ (d) $\dfrac{12}{25}$

    (e) $\dfrac{3}{500}$ (f) $\dfrac{3}{125}$

    (g) $\dfrac{69}{200}$ (h) $\dfrac{66}{125}$

Reinforcement

Extra Practice, Unit 7, Exercise 1, pp. 141-146

15. (a) $2\dfrac{3}{5}$ (b) $3\dfrac{1}{5}$

    (c) $1\dfrac{1}{4}$ (d) $6\dfrac{1}{20}$

Test

Tests, Unit 7, 1A and 1B, pp. 1-7

    (e) $3\dfrac{1}{500}$ (f) $2\dfrac{3}{40}$

    (g) $2\dfrac{51}{125}$ (h) $4\dfrac{1}{8}$

Chapter 2 – Approximation

Objectives

♦ Round a decimal to the nearest whole number.
♦ Round a decimal to 1 or 2 decimal places.

Notes

In *Primary Mathematics* 3A students learned to round numbers of up to 4 digits to the nearest ten, hundred, or thousand. In *Primary Mathematics* 4A students learned to round larger numbers to the nearest hundred, thousand, ten thousand, hundred thousand, and million. In *Primary Mathematics* 4B students learned to round decimals to a whole number and to one decimal place. In this chapter your student will review rounding decimals to a whole number or 1 decimal place and extend this to rounding decimals to 2 decimal places.

Being able to round numbers to a certain place can be used to approximate the answers to addition, subtraction, multiplication, or division problems. Quick approximations are useful with decimal numbers since it is easy to make a mistake in placing the decimal correctly in the answer. For example, the process for multiplying 42.4 x 0.7 is the same as for 424 x 7, but the answer is 29.68 instead of 2968. To check the placement of the decimal and if the answer makes sense we can round the numbers being multiplied to quickly get an approximate answer: 40 x 1 = 40 (or 40 x 0.7 = 28), so the answer cannot be 2.968 or 296.8. Rounding will also be used in division for decimal quotients. For example, 46 ÷ 7 = 6.6 to the nearest tenth (the actual answer is a non-terminating decimal).

In later levels, students will also be using estimates of irrational numbers, such as 3.14 for π, and will be rounding answers for the area and circumference of a circle. Decimal numbers are used in measurement, and later your student will learn in science that we also round the final answer in calcutions to a specific place since the answer from calculations cannot be any more accurate than the least accurate measurement.

Students have already used number lines to understand the process of rounding. With a number line, they can visually see what whole number, tenth, or hundredth the number is closest to. Number lines are also used in this chapter.

By convention, if a number is exactly halfway between two numbers it could be rounded to, it is rounded to the higher number. For example, 465 rounded to the nearest ten is 470.

To round a number to a specified place, we look at the digit in the next lower place. If it is 5 or greater than 5, we round up. If it is smaller than 5, we round down.

For example:

345.639 is 350 to the nearest ten.
345.639 is 346 to the nearest whole number.
345.639 is 345.6 to the nearest tenth, or 1 decimal place.
345.639 is 345.64 to the nearest hundredth, or 2 decimal places.

(1) Round decimals

Activity

Discuss instances where numbers are rounded. With whole numbers we often round populations or distances. For example, there are *about* 300 million people means there are 300 million people rounded to the closest hundred million (or possibly the closest million). A few hundred thousands one way or the other don't matter in the context of the discussion. The height of the tallest mountain is about 8800 meters above sea level. We don't really need to know its height to the nearest meter.

For a concrete example of rounding if needed, point out that each division on a meter stick is one thousandth of a meter (each centimeter is one hundredth of a meter). You can have your student measure something to the closest mark and write the measurement down as a decimal to 3 places. Then have him give the measurement to the nearest 2-decimal places, which would be the nearest centimeter mark. Each 10 centimeters is a tenth of a meter. Ask him to also give the measurement to the nearest 10 centimeter mark and write it down as a 1-place decimal.

Discussion

Concept pp. 13-14

Point out that the wavy equal sign stands for *about*.

In each of the examples, write the number and underline the digit in the place-value we are rounding to. Point out the next digit to the right and discuss how it affects the digit in the place we are rounding to in the rounded number. Tell your student that for this textbook and workbook, we will always round up if the next digit is 5 and there are no digits after. So 2.75 is about 2.8 when rounded to tenths.

Practice

Tasks 1-6, pp. 14-15

Workbook

Exercise 3, p. 7 (answers p. 11)

Reinforcement

Extra Practice, Unit 7, Exercise 2, pp. 147-148

Test

Tests, Unit 7, 2A and 2B, pp. 9-12

2.728 ≈ 3
↑
7 is more than 5, 2 ones becomes 3 ones.

2.728 ≈ 2.7
↑
2 is less than 5, 7 tenths stays 7 tenths.

2.728 ≈ 2.73
↑
8 is more than 5, 2 hundredths becomes 3 hundredths.

2.75 ≈ 2.8

---

1. (a) 7 m
   (b) 3 kg

2. (a) 2   (b) 3   (c) 19   (d) 9

3. 14.3
   14.4
   14.4

4. (a) 6.1   (b) 29.9   (c) 40.8   (d) 17.6

5. 3.15
   3.14
   3.15

6. (a) 5.17   (b) 8.04   (c) 10.81   (d) 23.72

Chapter 3 – Add and Subtract Decimals

Objectives

◆ Add decimal numbers.
◆ Subtract decimal numbers.
◆ Check the reasonableness of the sum or difference.

Material

◆ Place-value discs (0.001, 0.01, 0.1, 1, 10, and 100)
◆ Mental Math 1-2 (appendix)

Notes

Students learned to add and subtract decimals to 2 places in *Primary Mathematics* 2A using either mental math strategies or the standard algorithm. This concept is reviewed in this chapter and extended to decimals to 3 places.

In rewriting the problems vertically in order to use the standard algorithm, your student should take care to align the decimal. When discussing the steps in using any of the standard algorithms, use place-value names. For example, in using the standard algorithm for 9.79 + 4.86, we first add 9 hundredths and 6 hundredths to get 15 hundredths, which is the same as 1 tenth 5 hundredths. Do not say, "we first add 9 and 6."

Mental math strategies used with decimals are similar to strategies used with whole numbers. Your student needs to pay particular attention to place value, since with decimals the digit farthest to the right is not necessarily the smallest place in both numbers. For example, in 5.98 + 0.5, the 5 tenths needs to be added to 9 tenths, not 8 hundredths. If you are starting at this level and your he does not have the foundation in mental math strategies from earlier levels, have him use the standard algorithm. If you want him to use mental math strategies, he should probably learn how to use them with whole numbers first.

Students learned to round decimals to whole numbers to estimate answers to problems in *Primary Mathematics* 4B. Encourage your student to make a habit of estimating answers. Estimation helps reduce errors that may result from putting the decimal point in the wrong place and can help him to determine whether an answer is reasonable. Estimation is particularly useful in multiplication and division where there is more potential for error, particularly with the placement of the decimal point.

When using estimation to determine whether an answer is reasonable your student should round to numbers that allow her to find the estimated answer quickly. This could be to the same place in each number, such as 25.48 + 7.64 = 25 + 8, if she finds it easy to mentally add 25 and 8. But she could also round the numbers to 30 + 8, which is sufficient to check the reasonableness of the actual answer. Any estimates provided in this guide as answers are just possible estimations; your student might have a more or less precise estimate depending on how she rounded the numbers.

Depending on previous experience or retention of concepts, the lesson can take more than 1 day. The mental math pages do not all need to be done with the lesson but can be used later as review.

(1) Add or subtract decimals

Discussion

Tasks 1-2, p. 17

Your student should be able to do these problems mentally. The objective of these two tasks is to get the student to pay attention to the place value of the single non-zero digit being added to or subtracted from the multi-digit number. In these problems there is no renaming. Ask your student to actually write the answers to at least some, particularly 1(d) and 2(f-g).

> 1. (a) 4.936
>   (b) 4.539
>   (c) 4.516
>   (d) 4.53
>
> 2. (a) 2.665   (b) 3.242
>   (c) 1.276   (d) 4.527
>   (e) 3.129   (f) 2.041
>   (g) 6.106   (h) 5.2

Activity

Discuss the following problems.

6.275 + 0.04

Adding 4 hundredths to 7 hundredths will make a tenth. Add 27 hundredths and 4 hundredths.

> 6.$\underline{275}$ + 0.$\underline{04}$ = ?
> 27 + 4 = 31
>   /\
>   3 1
> 6.275 + 0.04 = 6.315

8.892 + 0.9

Add 1 and subtract a tenth.

> 8.892 + 0.9 = ?
> 8.892 + 0.9 = 8.892 + 1 − 0.1
>   = 9.792

6.274 − 0.007

There are not enough thousandths to subtract 7 thousandths. Subtract 7 thousandths from 74 thousandths.

> 6.$\underline{274}$ − 0.$\underline{007}$ = ?
>   74 − 7 = 3 + 64 = 67
>   /\
> 4 70
> 6.274 − 0.007 = 6.267

8.842 − 0.099

Subtract a tenth and add a thousandth.

> 8.842 − 0.099 = ?
> 8.842 − 0.099 = 8.842 − 0.1 + 0.001
>   = 8.743

Write the expressions at the right and have your student use mental math strategies to solve them. For example, in 5.555 + 0.005, since 55 and 5 is 60, then 0.055 and 0.005 is 0.06, so 5.5$\underline{55}$ + 0.00$\underline{5}$ = 5.5$\underline{6}$.

> 5.$\underline{555}$ + 0.00$\underline{5}$ (5.5$\underline{6}$)
> 5.$\underline{555}$ + 0.0$\underline{6}$ (5.$\underline{615}$)
> 5.$\underline{555}$ + 0.$\underline{7}$ ($\underline{6.255}$)
> 5.$\underline{555}$ − 0.00$\underline{8}$ (5.5$\underline{47}$)
> 5.$\underline{555}$ − 0.0$\underline{9}$ (5.$\underline{465}$)
> 5.$\underline{555}$ − 0.$\underline{5}$ (5.$\underline{055}$)

Discussion

Task 3, p. 17

Since there are a thousand thousandths in 1, we can "make 1" with thousandths in the same way as we "make 1000" with a 3-digit number.

Write the expression 4 − 0.456 and ask your student to solve it. She can simply subtract the decimal from one of the ones.

> 3. 0.544
>
> 1 = 0.900 + 0.090 + 0.010
> 0.456 + ?   0.400 + 0.500 = 0.900
>   0.050 + 0.040 = 0.090
>   0.006 + 0.004 = 0.010
> 0.456 + 0.544 = 1
>
> 4 − 0.456 = 3.544

Practice

Task 4, p. 17

> 4. (a) 0.54   (b) 0.69   (c) 0.931
>   (d) 0.931 (e) 1.58   (f) 0.889

See if your student can solve these without help. Provide guidance if needed. For example:

4(a) "Make 100" with 46 and write the answer as hundredths.

4(b) "Make 100" with 31 and write the answer as hundredths.

4(c) "Make 1000" with 69 and write the answer as thousandths.

4(d) The answer will be less than 1 and will be the same as the answer for 4(c).

4(e) Subtract 3 from 5 first to get 2, and then 0.42 from 2 (by subtracting from one of the ones).

4(f) The answer will be less than 1 since 11 is one less than 12, and can be solved using 1 − 0.111.

Discussion

Concept p. 16

> 6.47
> 1.74

Tell your student that whenever a problem seems difficult to do mentally, he can add or subtract using the standard algorithm. Rewrite the problem and go through the steps on paper. Relate them to the picture. If necessary use actual place-value discs. Point out that the digits need to be aligned carefully and if they are aligned correctly, the decimal point will also be aligned correctly.

Tasks 5-6, p. 17

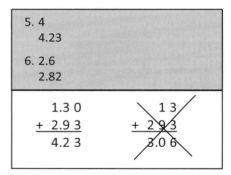

To find the actual values, have your student rewrite the problems vertically. Make sure she aligns the digits correctly. Add extra 0's on the end to have the both numbers go out to the same decimal place. Point out that estimating the answer is a way of checking that the digits in the correct places were added together.

Practice

Task 7, p. 17

> 7. (a) 6.15   (b) 14.81 (c) 69.38
>   (d) 1.89   (e) 5.11   (f) 57.22

Estimates can vary depending on how numbers were rounded.

Workbook

Exercise 4, p. 8 (answers p. 11)

Reinforcement

Mental Math 1-2

Extra Practice, Unit 7, Exercise 3, pp. 149-150

Test

Tests, Unit 7, 3A and 3B, pp. 13-16

Workbook

Exercise 1, p. 5

1. (a) 7.623
   (b) 4.605
   (c) 380.005
   (d) 8.002

2. (a) 5 ones, 0 tenths, 4 hundredths
   (b) 6 ones, 2 tenths, 3 hundredths, 8 thousandths

3. (a) 4
   (b) 7
   (c) 0.08
   (d) 0.004

4. (a) 0.248
   (b) 0.792
   (c) 3.78
   (d) 1.703
   (e) 0.085
   (f) 5.609

Exercise 2, p. 6

1. (a) <   (b) <
   (c) =   (d) <
   (e) >   (f) >

2. (a) 0.008, 0.009, 0.08, 0.09
   (b) 3.025, 3.205, 3.25, 3.502

3. (a) $\frac{2}{25}$ (b) $\frac{7}{50}$ (c) $\frac{29}{200}$

   (d) $\frac{51}{125}$ (e) $4\frac{253}{500}$ (f) $2\frac{3}{5}$

Exercise 3, p. 7

1. (a) 6
   (b) 21.5
   (c) 17.01

2. 0.08   2.31   4.08   3.26   1.80
   0.01   3.02   4.04   3.66   1.21

Exercise 4, p. 8

1. (a) 0.246   (b) 0.72
   (c) 3.78   (d) 10.054
   (e) 5.8   (f) 8.04
   (g) 14.864   (h) 2

2. Estimates can vary.
   (a) 48.598   (b) 6.369   (c) 40.066
   (d) 5.413   (e) 42.84   (f) 5.255

Chapter 4 – Multiply and Divide Decimals by a 1-Digit Whole Number

Objectives

◆ Multiply and divide a decimal number by a 1-digit whole number.
◆ Use estimation to check reasonableness of answers.
◆ Round the quotient to 1 or 2 decimal places.
◆ Express a fraction as a decimal correct to 2 decimal places.

Material

◆ Place-value discs (0.01, 0.1, 1, 10, and 100)
◆ Mental Math 3

Notes

In *Primary Mathematics* 4B students learned to multiply and divide a decimal number by a 1-digit whole number. This is briefly reviewed in this chapter. Students also learned to round the quotient to 1 decimal place. This is reviewed and extended to rounding the quotient to 2 decimal places. Your student will then learn to convert a fraction to a decimal using division where the quotient is a repeating or non-terminating decimal.

Your student will be asked to estimate the answer in order to check if his answer is reasonable. To estimate a product, we can round the decimal to one non-zero digit, e.g., 27.9 x 3 can be estimated by using 30 x 3 = 90 and 0.279 x 3 can be estimated using 0.3 x 3 = 0.9. In division, rather than rounding the dividend to the nearest whole number or tenth, we round to the closest multiple of the the divisor. For example, to estimate the answer for the division problem 31.2 ÷ 8, it would not be any easier to divide 30 by 8 than it is to divide 31.2 by 8, so it is not helpful to round the dividend to the nearest ten. 31.2 is close to 32 which is a multiple of 8, so 31.2 ÷ 8 is about 4. To estimate the answer for the division problem 3.12 ÷ 8, we can round the divisor to 3.2 and get the estimated answer of 0.4. Another valid estimate could be to simply think 3 ÷ 8 is $\frac{3}{8}$, 3 is less than half of 8, and so the answer will be slightly less than one half, 0.5. To estimate 3.84 ÷ 3, we could round 3.84 to 3 to give an estimated answer of 1, but we could also round it to 3.6 to give a more precise estimate of 1.2. An estimate is not an exact answer so don't insist on one correct estimate.

In *Primary Mathematics* 5A students learned to relate fractions to division and to convert an improper fraction (a fraction with a numerator equal to or larger than the denominator) into a mixed number or whole number using division. In this chapter, instead of writing the remainder as a fraction, your student will divide the fractional part and express it as a decimal correct to at most 2 decimal places. For example, instead of expressing the answer to 22 ÷ 7 as $3\frac{1}{7}$, your student will give the quotient as 3.14, correct to 2 decimal places.

Although it may be useful to teach your student the meaning of the terms dividend and divisor in order to use them in discussion, and they are used in this guide, being able to divide has nothing to do with these terms. He should know the term quotient from *Primary Mathematics* 3A, but if he is likely to mix up dividend with divisor, use terms such as the "whole" and "number of groups."

dividend ÷ divisor = quotient

$$\text{divisor)}\overline{\text{dividend}}^{\text{quotient}}$$

(1) Multiply, divide, and estimate

Discussion

Concept p. 18

| 9.06 |
| 1.35 |

Go through the steps for the multiplication and division problem. Don't just have your student look at the page, but have her actually do the problems on paper or whiteboard, step by step. When discussing the steps, always include the place value. For example, we are multiplying 5 tenths by 2 to get 10 tenths. Make sure she realizes why we have to add the hundredths place value to divide 8.1 by 6; in order to divide the remainder 3 tenths we need to rename it as 30 hundredths and the quotient of 30 hundredths divided by 6 is 5 hundredths. If she has trouble with the division, consider going back to an earlier level of *Primary Mathematics* to learn multiplication and division properly rather than trying to teach it here.

Tasks 1-2, p. 19

| 1. 6 |
| 6.24 |
| 2. 0.7 |
| 0.722 |

Have your student rewrite the problems and find the exact answer using the standard algorithm.

Task 1: To estimate 2.08 x 3, 2.08 is rounded to a whole number. Tell your student that to estimate in multiplication we usually round to the first non-zero digit. Ask him how he would round to estimate the answers to 20.8 x 3, 0.208 x 3, and 0.028 x 3. To help with placement of the decimal in the estimated answer, use place-value terms. 2 tenths x 3 is 6 tenths, 3 hundredths x 3 is 9 hundredths.

| $2.08 \times 3 \approx 2 \times 3 = 6$ |
| $20.8 \times 3 \approx 20 \times 3 = 60$ |
| $0.208 \times 3 \approx 0.2 \times 3 = 0.6$ |
| $0.028 \times 3 \approx 0.03 \times 3 = 0.09$ |

Task 2: To estimate 3.61 ÷ 5, 3.61 is rounded to 3.5. For division, we usually round the first two digits of the dividend (the whole) to a multiple of the divisor. 35 is the multiple of 5 that 36 is closest to. So we round 36 *tenths* to 35 tenths (3.5), to give an estimated answer of 7 *tenths*. In this case, we could have rounded to the whole number 4, since 4.0 ÷ 5 is easy to find. But 3.5 ÷ 5 is also easy to find, and gives a more precise estimate.

| $3.61 \div 5 \approx 3.5 \div 5 = 0.7$ |
| $36.1 \div 5 \approx 35 \div 5 = 7$ |
| $361 \div 5 \approx 350 \div 5 = 70$ |

Ask your student to estimate 3.61 ÷ 7. In this case, we cannot round to 4 and divide easily, so rounding to a whole number would be useless for an estimate. Then ask her to estimate 0.361 ÷ 7. We still round 36 to the nearest multiple of 7, but this time it is 35 hundredths, so the estimated answer is 5 hundredths.

| $3.61 \div 7 \approx 3.5 \div 7 = 0.5$ |
| $0.361 \div 7 \approx 0.35 \div 7 = 0.05$ |
| $0.036 \div 7 \approx 0.035 \div 7 = 0.005$ |

Practice

Task 3, p. 19

Workbook

Exercise 5, p. 9 (answers p. 18)

3. (a) 15	(b) 24	(c) 360
14.46	21.78	389.25
(d) 7	(e) 0.7	(f) 7
7.37	0.72	6.77

(2) Round the quotient

Activity

Write the problem 21.4 ÷ 7 and ask your student to solve it and find the quotient as a decimal. He should soon realize that there is always a remainder. You can have him carry out the division as far as he likes adding more places for ten thousandths, hundred thousandths and so on. Tell him that since there will never be a remainder of 0 in order to write the quotient as a decimal we eventually need to round it. Have him round the quotient to 1 and then to 2 decimal places. Point out that to round the quotient to 1 decimal place (tenths) we need to know the digit in the second decimal place (hundredths). Likewise, to round the quotient to 2 decimal places (hundredths) we need to know the digit in the third decimal place (thousandths).

```
21.4 ÷ 7

              3.0 5 7 1 4 2 ...
        7 ) 2 1.4 0 0 0 0 0
            2 1
            0 4 0
              3 5
              5 0
              4 9
              1 0
                7
                3 0
                2 8
                2 0
                1 4
                  6
```

Discussion

Tasks 4-5, pp. 19-20

```
4. (a) 2
   (b) 2.4

5. (a) 3
       3
   (b) 3.08
```

To give your student more practice with division, you can write the problem horizontally and have her rewrite and solve using the division algorithm, rather than just have her look at the text.

Reiterate that to round the quotient to 1 decimal place, we need to divide to 2 decimal places, and to round to 2 decimal places, we need to divide to 3 decimal places.

Practice

Task 6, p. 20

Workbook

Exercise 6, pp. 10-11
(answers p. 18)

```
6. (a)    0.0 8 5 ≈ 0.09    (b)    8.9 5 7 ≈ 8.96   (c)    1.2 0 6 ≈ 1.21
       9) 0.7 7                 7) 6 2.7                8) 9.6 5
          0.7 2                    5 6                     8
            5 0                    6 7                     1 6
            4 5                    6 3                     1 6
              5                    4 0                     0 5 0
                                   3 5                       4 8
                                   5 0                         2
                                   4 9
                                     1

   (d)    6.9 1 8 ≈ 6.92    (e)    6.9 2 2 ≈ 6.92   (f)    6.9 7 6 ≈ 6.98
       6) 4 1.5 1              4) 2 7.6 9              3) 2 0.9 3
          3 6                    2 4                     1 8
          5 5                    3 6                     2 9
          5 4                    3 6                     2 7
            1 1                    0 9                     2 3
              6                    8                       2 1
              5 0                  1 0                     2 0
              4 8                    8                     1 8
                2                    2                       2
```

(3) Divide fractions

Discussion

Tasks 7-9, p. 20

Task 7: This shows two methods for converting a fraction to a decimal. The first involves finding an equivalent fraction with a denominator of 10, 100, or 1000, and the second involves relating fractions to division.

Task 8: Have your student use both methods. 8 is neither a factor of 10 nor of 100, but it is a factor of 1000.

Task 9: Again have your student use both methods. In this case, he would use a denominator of 10, since 5 is a factor of 10. Be sure he realizes that we only need to find the decimal equivalent of the fractional part of the mixed number.

Discuss which method is easier. If it is easy to recognize the denominator as a factor of 10, 100, or 1000, and know the corresponding factor, the first method can be easier to do mentally.

> 7. 0.75   0.75
>
> 8. 0.125
>
> 9. 3.4

Practice

Task 10, p. 20

> 10. (a) 2.25   (b) 4.375   (c) 1.8   (d) 6.875

Discussion

Tasks 11-12, 14, p. 21

Task 11: To arrange a set of numbers that consists of both fractions and decimals in order, either the fraction needs to be renamed as a decimal or the decimals renamed as fractions. Ask your student which is easier. Generally, renaming fractions as decimals is easier, since even if we rename decimals as fractions we still may need to find equivalent fractions. It is usually easier to compare decimals.

Task 12(a): In this example, 9 is not a factor of 10, 100, or 1000, so to convert to a decimal we need to divide. You can have your student carry on the division to realize that there will always be a remainder of 5. So we cannot give an exact decimal value for $\frac{5}{9}$. We can either say that it is about 0.6, or it equals 0.6, *correct to 1 decimal place*. Saying that the answer is correct to 1 decimal place means that the exact answer is greater than or equal to 0.55 and less than 0.65.

> 11. $\frac{5}{8}$ = 0.625
>
> 0.6, $\frac{5}{8}$, 0.652, 2
>
> 12. (a) 0.6
>
> (b) 3.6
>
> 14. 0.67
>
> 4.67
>
> 4.67
>
> 4.67

Task 14: Here we are finding the decimal equivalent correct to 2 decimal places. The farther out we carry the division, the more accurate the answer.

Write the set of numbers at the right and ask your student to put them in increasing order. She may simply start dividing the fractions to find the decimals. After she is done, discuss ways to put the numbers in order without doing any calculations. Obviously, 1.67 is the greatest. $\frac{3}{7}$ is smallest because it is less than $\frac{1}{2}$. $\frac{7}{9}$ is close to 1.

> 1.67, $\frac{3}{7}$,   0.546, $\frac{7}{9}$
>
> $\frac{3}{7}$, 0.546, $\frac{7}{9}$, 1.67

Practice

Tasks 13 and 15, p. 21

Workbook

Exercise 7, p. 12 (answers p. 18)

13.	(a) 0.8	(b) 0.6	(c) 0.4	(d) 0.8
	(e) 2.7	(f) 4.9	(g) 3.2	(h) 1.9
15.	(a) 0.43	(b) 0.63	(c) 0.22	(d) 0.17
	(e) 5.78	(f) 1.33	(g) 4.71	(h) 8.38

Reinforcement

Your student should eventually become familiar with the decimal representation of some common fractions, if she is not already.

$\frac{1}{2} = 0.5$ $\frac{1}{3} = 0.333...$ $\frac{2}{3} = 0.666...$

$\frac{1}{4} = 0.25$ $\frac{3}{4} = 3 \times \frac{1}{4} = 0.75$

$\frac{1}{5} = 0.2$ $\frac{2}{5} = 2 \times 0.2 = 0.4$ $\frac{3}{5} = 3 \times 0.2 = 0.6$ $\frac{4}{5} = 4 \times 0.2 = 0.8$

$\frac{1}{6} = 0.1666...$ $\frac{1}{8} = 0.125$

Enrichment

Mental Math 3
(Give answers as decimals of up to 3 decimal places.)

The fractions in this section are given as mixed numbers with the fractional part in its simplest form. Ask your student how he would find the quotient for a problem like 210 ÷ 56. Point out that this one would not be much fun to divide. In some cases, where the divisor and dividend have common factors, we can first simplify, and then divide. It is easy to tell if they have common factors of 2, 3, or 5.

$$210 \div 56 = \frac{210}{56} = \frac{105}{28} = \frac{15}{4} = 3\frac{3}{4} = 3.75$$

(4) Practice

Practice

Practice A, p. 22

Problem 4: It is not always necessary to completely convert the fraction to a decimal in these types of problems.

For example, in 4(d), $\frac{3}{8}$ is greater than $\frac{2}{8}$ which is $\frac{1}{4}$ which is greater than 0.065.

Problems 8(a-b): Your student might try to round the numbers first and then add or subtract. Even if she does not, you can ask her to try it with 8(a) and see what answer she gets (10.35 and 8.32). These are not correct answers because we cannot say that they are correct to 2 decimal places. If a problem asks for an answer to a given number of decimal places for addition, subtraction, or multiplication problems, not just division problems, we need to round after performing the calculations, not before.

1. (a) 0.6	(b) 6	(c) 0.06	(d) 0.006
2. (a) 5.509			
(b) 2.819			
(c) 13.51			
3. (a) 3.007	(b) 3.8	(c) 6.25	(d) 8.125
4. (a) =	(b) >		
(c) <	(d) >		
(e) =	(f) >		
(g) =	(h) >		
5. (a) 0.12	(b) 7.51	(c) 40.08	(d) 81.14
(e) 0.73	(f) 3.12	(g) 59.01	(h) 18.61
6. (a) 6.27 km	(b) 4.08 kg	(c) 0.19 ℓ	(d) 20.25 ℓ
7. (a) 0.13	(b) 0.57	(c) 2.56	(d) 5.67
8. (a) 10.34	(b) 8.29		
(c) 6.02	(d) 4.69		

Reinforcement

Extra Practice, Unit 7, Exercise 4, pp. 151-152

Tests

Tests, Unit 7, 4A and 4B, pp. 17-20

Enrichment

If your student did not already do the similar activity in the guide for *Primary Mathematics* 4, ask him to express the fractions $\frac{1}{9}, \frac{2}{9}, \frac{3}{9}, \frac{4}{9}, \frac{5}{9}, \frac{6}{9}, \frac{7}{9}$, and $\frac{8}{9}$ as decimals. He should see the pattern pretty quickly. Ask him, if he continued the pattern, what he would get for $\frac{9}{9}$, and whether the never-ending decimal 0.9999... is equivalent to 1. (Mathematically, 0.9999... is 1. If you multiply both sides of $\frac{1}{3}$ = 0.3333... by 3, you get 1 = 0.9999...).

Workbook

Exercise 5, p. 9

1. (a) 6   (b) 20   (c) 56
     5.7   19.1   56.21
   (d) 0.16   (e) 15   (f) 80
     0.184   13.521   82.3
   (g) 12   (h) 0.6   (i) 15
     12.8   0.62   15.5
   (j) 4   (k) 4   (l) 9
     3.7   3.75   8.9

Exercise 6, pp. 10-11

1. (a)
```
        7.7 7 ≈ 7.8
    9)7 0.0 0
      6 3
        7 0
        6 3
          7 0
          6 3
            7
```
   (b)
```
        4.5 0 ≈ 4.5
    4)1 8.0 1
      1 6
        2 0
        2 0
          0 1
```

2. (a)
```
        4.3 3 6 ≈ 4.34
    5)2 1.6 8 0
      2 0
        1 6
        1 5
          1 8
          1 5
            3 0
            3 0
              0
```
   (b)
```
        6.9 2 1 ≈ 6.92
    6)4 1.5 3 0
      3 6
        5 5
        5 4
          1 3
          1 2
            1 0
              6
              4
```

   (c)
```
        0.0 7 5 ≈ 0.08
    7)0.5 3 0
      4 9
        4 0
        3 5
          5
```
   (d)
```
        3.0 0 6 ≈ 3.01
    8)2 4.0 5 0
      2 4
        0 0 5 0
            4 8
            2
```

3.

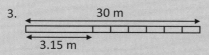

30 m − 3.15 m = 26.85 m

26.85 m ÷ 6 = 4.475 m ≈ **4.48**

She used about 4.48 m of raffia for each pot holder.

4.

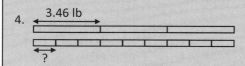

3.46 lb ÷ 3 ≈ **1.15 lb**

Each bag had about 1.15 lb of nuts.

Exercise 7, p.12

1. (a)
```
        0.8 8 ≈ 0.9
    9)8.0 0
      7 2
        8 0
        7 2
          8
```
   $\frac{8}{9}$ ≈ **0.9**

   (b)
```
        0.1 6 ≈ 0.2
    6)1.0 0
      6
      4 0
      3 6
        4
```
   $4\frac{1}{6}$ ≈ **4.2**

2. (a)
```
        0.6 6 6 ≈ 0.67
    3)2.0 0 0
      1 8
        2 0
        1 8
          2 0
          1 8
            2
```
   $\frac{2}{3}$ ≈ **0.67**

   (b)
```
        0.4 2 8 ≈ 0.43
    7)3.0 0 0
      2 8
        2 0
        1 4
          6 0
          5 6
            4
```
   $\frac{3}{7}$ ≈ **0.43**

   (c)
```
        0.6 2 5 ≈ 0.63
    8)5.0 0 0
      4 8
        2 0
        1 6
          4 0
          4 0
            0
```
   $5\frac{5}{8}$ ≈ **5.63**

   (d)
```
        0.1 4 2 ≈ 0.14
    7)1.0 0 0
      7
      3 0
      2 8
        2 0
        1 4
          6
```
   $9\frac{1}{7}$ ≈ **9.14**

Chapter 5 – Multiplication by Tens, Hundreds or Thousands

Objectives

♦ Multiply a decimal by tens, hundreds or thousands.

Material

♦ Place-value discs (0.01, 0.1, 1, 10, and 100)
♦ Mental Math 4

Notes

In *Primary Mathematics* 4B students learned to multiply a decimal by a 1-digit whole number. In *Primary Mathematics* 5A they learned to multiply a whole number by tens, hundreds, or thousands. In this chapter the concepts are extended to multiplying decimals by tens, hundreds, or thousands.

When a whole number is multiplied by 10, a 0 is added to the whole number. When a decimal is multiplied by 10, the decimal point moves one place to the right. This is essentially what happens with a whole number as well, since the decimal is at the right of the ones place. In both cases, the digit in each place value of the product is now ten times as much as the same digit in the original number. It has "moved" one place to the left relative to the decimal point.

Similarly, when either a whole number or a decimal is multiplied by 100 or 1000, each digit is 100 or 1000 times greater, moving it 2 or 3 places to the left, respectively. This results in the decimal point moving 2 or 3 places to the right.

$$1234. \times 10 = 12,340$$
$$1.234 \times 10 = 12.34$$
$$1234 \times 100 = 123,400$$
$$1.234 \times 100 = 123.4$$
$$1.2 \times 100 = 120$$
$$1234 \times 1000 = 1,234,000$$
$$1.234 \times 1000 = 1234$$

We can multiply by tens, hundreds, or thousands in two steps, first multiplying by the non-zero digit in the tens place, and then multiplying by 10, 100, or 1000. The steps can be reversed. From working with factors in both *Primary Mathematics* 4A and 5A, your student should know that multiplying a number by its factors in any order is the same as multiplying by that number. For example, 25 x 16 = 25 x 4 x 4 = 100 x 4 = 400.

$$
\begin{aligned}
4.3 \times 20 &= 4.3 \times 2 \times 10 \\
&= 8.6 \times 10 \\
&= 86
\end{aligned}
$$

$$
\begin{aligned}
4.3 \times 20 &= 4.3 \times 10 \times 2 \\
&= 43 \times 2 \\
&= 86
\end{aligned}
$$

$$
\begin{aligned}
3.421 \times 3000 &= 3.421 \times 3 \times 1000 \\
&= 10.263 \times 1000 \\
&= 10,263
\end{aligned}
$$

The textbook uses pictures of place-value discs and place-value charts as a visual model. Use your judgment on whether to use actual place value disks. For some students physically trading discs or writing the digits on a place value chart before and after multiplying and looking at their relative positions is more effective in illustrating the concept than simply looking at static pictures in a textbook. For others, the pictures are sufficient. Be sure to give your student the best opportunity to understand the concept even if it involves pulling out simple manipulatives.

(1) Multiply a decimal by tens

Discussion

Concept p. 23

Let your student use actual number discs if it helps. Each tenth, when multiplied by 10, is 1. Each hundredth, when multiplied by 10, is a tenth. Each thousandth, when multiplied by 10, is a hundredth. In each case, ask your student which place the digit 8 is in both before and after multiplying by 10. The digit "moves" one place to the left by becoming ten times as much. Then ask her what that does to the decimal point. It makes it "move" one place to the right.

Practice

Task 1, p. 24

Discussion

Tasks 2-3, p. 24

Let your student use actual number discs or write numbers on a place-value chart if it helps.

Task 2: To multiply 3.42 by 10, each digit is multiplied by 10. 3 ones becomes 3 tens, 4 tenths becomes 4 ones, and 2 hundredths becomes 2 tenths.

Task 3: This task shows what happens to each digit on a place-value chart and emphasizes that the position of each digit, when multiplied by 10, is one place value to the left of its original position. The end result is that the decimal point "moves" one place to the right.

Practice

Task 4, p. 24

Discussion

Task 5, p. 25

To multiply by 4 tens, we can multiply first by 4 and then by 10, since 40 is 4 x 10. Your student can use the multiplication algorithm for the first step.

Practice

Task 6, p. 25

Workbook

Exercise 8, pp. 13 (answers p. 26)

---

1 tenth   x 10 = 1 one so
8 tenths x 10 = 8 ones
0.8   x 10 = 8
↑   ↑
Tenths place   Ones place

1 hundredth   x 10 = 1 tenth so
8 hundredths x 10 = 8 tenths
0.08   x 10 = 0.8
↑   ↑
Hundredths place Tenths place

1 thousandth   x 10 = 1 hundredth so
8 thousandths x 10 = 8 hundredths
0.008   x 10 = 0.08
↑   ↑
Thousandths place   Hundredths place

---

1. (a) 6   (b) 8   (c) 9
   (d) 0.2   (e) 0.4   (f) 0.3
   (g) 0.05   (h) 0.06   (i) 0.07

---

2. 34.2

3. 0.35

---

3   x 10 = 30
0.4   x 10 =   4
0.02   x 10 =   0.2
3.42   x 10 = 34.2

0.0 3 5 x 1 0 = 0.3 5

---

4. (a) 1.2   (b) 0.68   (c) 3.45
   (d) 20.5   (e) 32.1   (f) 14.39
   (g) 75   (h) 104   (i) 118

---

5. 21.2

0.53 x 40 = 0.53 x 4 x 10
           = 2.12 x 10
           = 21.2

---

6. (a) 0.18   (b) 3.2   (c) 45
   (d) 6.4   (e) 476.7   (f) 194.88

(2) Multiply a decimal by hundreds or thousands

Discussion

Tasks 7-10, pp. 25-26

The same concepts apply to multiplying a decimal by 100 or 1000 as multiplying a decimal by 10.

Tasks 7-8: When each digit in a number is multiplied by 100, its position in the product will be two places to the left of its original position; a digit that was in the thousandths place is now in the tenths place, a digit that was in the hundredths place is now in the ones place, and so on. This has the effect of moving the decimal point two places to the right.

Task 9-10: Similarly, multiplying a number by 1000 has the effect of moving the decimal point three places to the right.

After your student has finished Task 10, ask him to multiply 0.054 by 6000 rather than 1000. The product will now be 6 times more, and can be found either by multiplying the decimal first by 6 and then 1000, or first by 1000 and then by 6. Your student might find the second method easier, since 54 has fewer decimal places to keep track of than 0.054 when multiplying by 6.

Let your student use the multiplication algorithm when he wants to.

If the problem can be solved mentally, it is usually easier to keep track of the place values by moving the decimal point first, then multiply by the whole number. You can have your student try it both ways with some multiplication problems that are easy to solve mentally.

Practice

Tasks 11-14, p. 26

Workbook

Exercise 9, pp. 14-15 (answers p. 26)

Reinforcement

Extra Practice, Unit 7, Exercise 5, pp. 153-154

Mental Math 4

Test

Tests, Unit 7, 5A and 5B, pp. 21-24

---

7. 0.7

8. 423

9. 6

10. 54

$\_\_ 4.2\ 3 \times 1\ 0\ 0 = 4\ 2\ 3$

$\_\_ 0.0\ 3\ 5 \times 1\ 0\ 0\ 0 = 3\ 5$

$0.054 \times 6000 = 0.054 \times 6 \times 1000$
$= 0.324 \times 1000$
$= 324$

$0.054 \times 6000 = 0.054 \times 1000 \times 6$
$= 54 \times 6$
$= 324$

$$\begin{array}{r} 0.0\ 5\ 4 \\ \times \quad\ 6 \\ \hline \end{array} \qquad \begin{array}{r} 5\ 4 \\ \times\ 6 \\ \hline \end{array}$$

$0.06 \times 500 = 6 \times 5 = 30$
$0.007 \times 90 = 0.07 \times 9 = 0.63$
$0.34 \times 2000 = 340 \times 2 = 680$
$1.4 \times 30 = 14 \times 3 = 42$

---

11. (a) 0.3   (b) 320   (c) 132.5
    (d) 90   (e) 3620   (f) 13,400

12. 840.6

13. 8406

14. (a) 2.4   (b) 72   (c) 616
    (d) 150   (e) 1500   (f) 20,480

Chapter 6 – Division by Tens, Hundreds or Thousands

Objectives

♦ Divide a decimal by tens, hundreds, or thousands.

Material

♦ Place-value discs (0.01, 0.1, 1, 10, and 100)
♦ Mental Math 5-6

Notes

In *Primary Mathematics 4A* students learned to divide tens, hundreds, or thousands by tens, hundreds, or thousands. In this chapter the concepts are extended to dividing decimals by tens, hundreds, or thousands.

When the digit in any place in a whole number or decimal is divided by 10, its value becomes one tenth as much, and its position in the quotient is one place to the right of its original position. This has the effect of moving the decimal point one place to the left. Students have found earlier that when a whole number with a 0 in the ones place is divided by 10, a 0 is removed from the whole number. This is because in the quotient the 0 ends up on the other side of the decimal and so is dropped.

$$12{,}340. \div 10 = 1{,}234$$
$$1234.5 \div 10 = 123.45$$
$$0.2 \div 10 = 0.02$$

Similarly, when a decimal or whole number is divided by 100, the digit in any place becomes one hundredth as much and its position in the quotient is two places to the right of its original position. The decimal point is therefore two places to the left. The same thing occurs when a number is divided by 1000, except that the digits are now three places to the right, and the decimal point therefore three places to the left.

$$12{,}300. \div 100 = 123$$
$$1234.5 \div 100 = 12.345$$
$$1.2 \div 100 = 0.012$$

$$12{,}000 \div 1000 = 12$$
$$123 \div 1000 = 0.123$$
$$2 \div 1000 = 0.002$$

To divide a number by a multiple of 10, 100, or 1000, we can first divide by a whole number, and then by 10, 100, or 1000, or the other way around. It is probably easier for your student to divide by the 1-digit whole number first, since there are fewer decimal places to deal with.

Again, the textbook shows place-value disks and then place-value charts to illustrate what happens to each digit when it is divided by 10, 100, or 1000. You may want to introduce each new concept by having your student physically trade discs in and replace each one by one a tenth, hundredth, or thousandth time as much, or writing the digits on the place value chart before and after dividing and looking at their relative positions.

$$24 \div 60 = 24 \div 6 \div 10$$
$$= 4 \div 10$$
$$= 0.4$$
$$24 \div 60 = 24 \div 10 \div 6$$
$$= 2.4 \div 6$$
$$= 0.4$$

$$31.2 \div 600 = 31.2 \div 6 \div 100$$
$$= 5.2 \div 100$$
$$= 0.052$$
$$31.2 \div 600 = 31.2 \div 100 \div 6$$
$$= 0.312 \div 6$$
$$= 0.052$$

(1) Divide a decimal by tens

Discussion

Concept p. 27

Let your student use actual number discs if it helps. Each one, when divided by 10, becomes a tenth. Each tenth, when divided by 10, becomes a hundredth. Each hundredth, when divided by 10, becomes a thousandth. In each case, ask your student which place the digit 3 is in both before and after multiplying by 10. The digit "moves" one place to the right by becoming a tenth as much. Then ask her what that does to the decimal point. It makes it "move" one place in the other direction.

> 1 one ÷ 10 = 1 tenth so
> 3 ones ÷ 10 = 3 tenths
> 3   ÷ 10 = 0.3
> ↑   ↑
> Ones place   Tenths place
>
> 1 tenth ÷ 10 = 1 hundredth so
> 3 tenths ÷ 10 = 3 hundredths
> 0.3   ÷ 10 = 0.03
> ↑   ↑
> Tenths place   Hundredths place
>
> 1 hundredth ÷ 10 = 1 thousandth so
> 3 hundredths ÷ 10 = 3 thousandths
> 0.03   ÷ 10 = 0.003
> ↑   ↑
> Hundredths place   Thousandths place

Practice

Task 1, p. 28

> 1. (a) 0.8   (b) 0.08   (c) 0.008
>   (d) 0.2   (e) 0.02   (f) 0.002
>   (g) 0.6   (h) 0.06   (i) 0.006

Discussion

Tasks 2-3, p. 28

Task 2: To divide 0.46 by 10, each digit is divided by 10. 4 tenths becomes 4 hundredths and 6 hundredths becomes 6 thousandths.

Task 3: This tasks shows what happens to each digit on a place-value chart and emphasizes that the position of each digit, after being divided by 10, is one place value to the right of its original position. The end result is that the decimal point "moves" one place to the left.

Write the expression 530 ÷ 10 and ask your student to find the quotient. 53 tens divided by ten is 53. Your student may remember that an easy way to divide a whole number that ends in 0 by 10 is simply to remove the 0. Ask him whether this follows the same pattern; dividing by 10 moves the decimal point one place to the left. In a whole number, we don't write the decimal point, because there are no digits in any of the decimal places, but if there were, it would be right after the 0. If we move the decimal point one place to the left in 530.0, we get 53.00, which we write as 53.

> 2. 0.046
>
> 3. 0.53
>
> 0.4   ÷ 10 = 0.04
> 0.06 ÷ 10 = 0.006
> 0.46 ÷ 10 = 0.046
>
> 5.3 ÷ 1 0 = 0.5 3

> 530 ÷ 10
>
> 5 3 0.0 ÷ 1 0 = 5 3.00 = 53

Practice

Task 4, p. 28

Activity

Write the expression 31.2 ÷ 40 and discuss its solution. You can tell your student that we can solve this by first dividing by 10 and then by 4, or by 4 and then by 10. Division by 4 can be done with the division algorithm or mentally. Ask her whether she thinks it is easier to divide by 10 first or by 4 first. Dividing by 4 first may be easier since there are fewer decimal places to worry about. Remind her that a quick estimate can help with errors in placement of the decimal or forgetting to do the second step. Because 40 is slightly larger than 31, the answer will be a bit less than 1.

Though your student should have learned that we can solve a multiplication expression by splitting the number into factors and multiplying in any order, he has not explicitly been taught that we can split the divisor into factors and divide by those factors in any order. You may want to explain that although we don't usually write fractions with decimals in the numerator and denominator, this division can still be represented by the fraction $\frac{31.2}{40}$. We are essentially finding an equivalent fraction with 1 in the denominator. We can divide both the numerator and denominator first by 10 and then by 4, or first by 4 and then by 10. Dividing the numerator and denominator first by 10 and then by 4 has the effect of first moving the decimal point over one place to the left in the numerator and denominator and then dividing by 4.

Make sure your student understands that because division is related to fractions, we can solve the multiplication problem by moving the decimal in both the numerator and denominator the same way to get an equivalent fraction. This is different than multiplication. 31.2 x 40 is not the same as 3.12 x 4.

Practice

Tasks 5-6, p. 29

Workbook

Exercise 10, p. 16 (answers p. 26)

4. (a) 0.023   (b) 0.045   (c) 0.012
   (d) 0.25   (e) 0.68   (f) 0.53
   (g) 1.2   (h) 3.9   (i) 10.3

$31.2 ÷ 40 = 31.2 ÷ 10 ÷ 4$
$= 3.12 ÷ 4$
$= 0.78$

$$\begin{array}{r} 0.7\,8 \\ 4\overline{)3.1\,2} \\ \underline{2\,8} \\ 3\,2 \\ \underline{3\,2} \end{array}$$

$31.2 ÷ 40 = 31.2 ÷ 4 ÷ 10$
$= 7.8 ÷ 10$
$= 0.78$

$$\begin{array}{r} 7.8 \\ 4\overline{)3\,1.2} \\ \underline{2\,8} \\ 3\,2 \\ \underline{3\,2} \end{array}$$

$$\frac{31.2}{40} = \frac{7.8}{10} = \frac{0.78}{1} = 0.78$$

$$\frac{31.2}{40} = \frac{3.12}{4} = \frac{0.78}{1} = 0.78$$

$$4.0\overline{)3.1.2}$$

$$\begin{array}{r} 0.7\,8 \\ 4.0\overline{)3.1\,2} \\ \underline{2\,8} \\ 3\,2 \\ \underline{3\,2} \end{array}$$

5. 0.07

6. (a) 0.2   (b) 0.2   (c) 0.7
   (d) 0.08   (e) 0.017   (f) 0.043

(2) Divide a decimal by hundreds or thousands

Discussion

Tasks 7-10, pp. 25-26

Tasks 7-8: When each digit in a number is divided by 100, its position in the product will be two places to the right of its original position; a digit that was in the ones place is now in the hundredths place, a digit that was in the tenths place is now in the thousandths place, and so on. This has the effect of moving the decimal point two places to the left.

Tasks 9-10: Similarly, dividing a number by 1000 has the effect of moving the decimal point three places to the left.

After your student has finished Task 10, ask him to divide 31.2 by 600. As with division by tens, we can first divide by 6 and then by 100, or first by 100 and then by 6. Division by 6 can be done with the standard algorithm.

Optional: You can show your student another method in which we write the division algorithm and move the decimal point over for both dividend and divisor (numerator and denominator) until the divisor is a whole number. We can do this because 31.2 ÷ 600 is the same as $\frac{31.2}{600}$ and we are essentially finding an equivalent fraction with a denominator of 1 in two steps, first dividing numerator and denominator by 100 and then by 6.

Write a few problems that your student should be able to do mentally. For mental math, it is usually easier to divide by the 1-digit whole number first, and then move the decimal point.

Practice

Tasks 11-14, p. 30

Workbook

Exercise 11, pp. 17-18 (answers p. 26)

Reinforcement

Extra Practice, Unit 7, Exercise 6, pp. 155-156

Mental Math 5-6

Test

Tests, Unit 7, 6A and 6B, pp. 25-28

---

7. 0.04

8. 0.528

9. 0.005

10. 0.062

$52.8 \_ \_ \div 100 = 0.5\ 2\ 8$

$62 \_ \_ \_ \div 1000 = 0.0\ 6\ 2$

$31.2 \div 600 = 31.2 \div 100 \div 6$
$= 0.312 \div 6$
$= 0.052$

$31.2 \div 600 = 31.2 \div 6 \div 100$
$= 5.2 \div 100$
$= 0.052$

```
    5.2
6)3 1.2
  3 0
    1 2
    1 2
```

```
6,0 0)3 1.2        6.0 0).0 5 2
                        3 1 2
                        3 0
                          1 2
                          1 2
```

$63 \div 900 = 7 \div 100 = 0.07$
$68 \div 2000 = 34 \div 1000 = 0.034$
$270 \div 300 = 90 \div 100 = 0.9$
$6 \div 500 = 1.2 \div 100 = 0.012$

---

11. (a) 0.08   (b) 0.9   (c) 0.015
    (d) 0.004   (e) 0.2   (f) 0.324

12. 0.23

13. 0.023

14. (a) 0.004 (b) 0.004 (c) 0.016
    (d) 0.002 (e) 0.013 (f) 0.102

Workbook

Exercise 8, p. 13

1. (a) 0.3   (b) 0.09
   (c) 0.67   (d) 8.4
   (e) 29   (f) 3.21
   (g) 52.4   (h) 354
   (i) 60.15   (j) 4128

2. (a) 0.09 x 20 = 0.18 x 10 = **1.8**
   Or
   0.09 x 20 = 0.9 x 2 = 1.8
   (b) 128
   (c) 277.8
   (d) 1832
   (e) 1116

Exercise 9, pp. 14-15

1.

Number	x 10	x 100	x 1,000
0.324	**3.24**	**32.4**	**324**
1.635	**16.35**	**163.5**	**1635**
3.004	**30.04**	**300.4**	**3004**
8.19	**81.9**	**819**	**8190**
20.4	**204**	**2040**	**20,400**

2. (a) 616.6   (b) 200.9
   (c) 520.1   (d) 306.5
   (e) 72   (f) 8625
   (g) 4860   (h) 3700

3. (a) 10   (b) 10
   (c) 100   (d) 1000
   (e) 10   (f) 1000
   (g) 1000   (h) 100
   (i) 100   (j) 1000

4. (a) 0.06 x 200 = 0.12 x 100 = **12**
   Or
   0.06 x 200 = 6 x 2 = 12
   (b) 102
   (c) 2720
   (d) 1560
   (e) 387,000
   (f) 224,560
   (g) 76,320
   (h) 29,160

Exercise 10, p. 16

1. (a) 0.6   (b) 0.03
   (c) 0.005   (d) 0.034
   (e) 0.12   (f) 1.9
   (g) 2.05   (h) 0.365
   (i) 23.9   (j) 0.058

2. (a) 0.8 ÷ 20 = 0.4 ÷ 10 = **0.04**
   Or
   0.8 ÷ 20 = 0.08 ÷ 2 = 0.04
   (b) 0.074
   (c) 0.089
   (d) 0.912
   (e) 0.423

Exercise 11, pp. 17-18

1.

Number	÷ 10	÷ 100	÷ 1,000
203	**20.3**	**2.03**	**0.203**
8	**0.8**	**0.08**	**0.008**
7050	**705**	**70.5**	**7.05**
58	**5.8**	**0.58**	**0.058**
1458	**145.8**	**14.58**	**1.458**

2. (a) 0.54   (b) 0.203
   (c) 28.2   (d) 0.034
   (e) 4.525   (f) 3.4
   (g) 0.073   (h) 0.002

3. (a) 10   (b) 100
   (c) 1000   (d) 100
   (e) 1000   (f) 10
   (g) 100   (h) 1000
   (i) 10   (j) 1000

4. (a) 7.2 ÷ 200 = 3.6 ÷ 100 = **0.036**
   Or
   7.2 ÷ 200 = 0.072 ÷ 2 = 0.036
   (b) 0.03
   (c) 0.106
   (d) 0.072
   (e) 0.003
   (f) 0.013
   (g) 0.098
   (h) 0.121

Chapter 7 – Multiplication by a 2-Digit Whole Number

Objectives

◆ Multiply a decimal by a 2-digit whole number.
◆ Estimate the product in multiplication of decimals.

Notes

In *Primary Mathematics* 4A students learned to multiply a whole number by a 2-digit whole number. This was reviewed in *Primary Mathematics* 5A. Here, the skill is extended to multiplication of a decimal by a 2-digit whole number. If your student cannot multiply a whole number by a 2-digit whole number using the standard algorithm, you should consider going back to *Primary Mathematics* 4A where it is first taught so that he thoroughly understands the renaming process involved.

Multiplication by 2 digits involves multiplying by the tens and multiplying by the ones and then adding the two partial products. The multiplication algorithm is a short-cut where the problem is written vertically so that the places can be aligned which makes it easier to keep track of the place values of the digits. It also speeds up the computation involved with multiplying by a singe digit (e.g., 567 x 2 = 14 + 120 + 1000) since renamed digits in the same place can be added immediately rather than later. Traditionally, when multiplying using the vertical representation, the first number is multiplied by the ones in the 2-digit number first, and then by the tens. This helps students align the columns correctly by placing or imagining a 0 below the ones of the first partial product to indicate that the answer is tens. 5 *tens* x 567 = 2835 *tens*.

$$567 \times 52 \approx 600 \times 50 = 30{,}000$$

$$
\begin{aligned}
567 \times 52 &= (567 \times 50) + (567 \times 2) \\
&= (567 \times 5 \times 10) + (567 \times 2) \\
&= (2835 \times 10) + 1134 \\
&= 28{,}350 + 1134 \\
&= 29{,}484
\end{aligned}
$$

```
      5 6 7
  x      5 2
      1 1 3 4   =  567 x 2
    2 8 3 5 0   =  567 x 50
    2 9 4 8 4
```

Multiplication of a decimal by a 2-digit whole number is a similar process to multiplication of a whole number. Usually the decimal is not written in the partial products but is added only after they have been summed. The digits need to be aligned correctly. The decimal point in the final answer can then be aligned with the decimal point in the decimal being multiplied. In the example at the right, one way to think of the process is that 567 hundredths is being multiplied by 52, so the answer is 29,484 hundredths, and so the decimal goes two places in from the right.

$$5.67 \times 52 \approx 6 \times 50 = 300$$

$$
\begin{aligned}
5.67 \times 52 &= (5.67 \times 50) + (5.67 \times 2) \\
&= (5.67 \times 5 \times 10) + (5.67 \times 2) \\
&= (28.35 \times 10) + 11.34 \\
&= 283.50 + 11.34 \\
&= 294.84
\end{aligned}
$$

```
      5.6 7
  x      5 2
      1 1 3 4   =  5.67 x 2
    2 8 3 5 0   =  5.67 x 50
    2 9 4.8 4
```

Estimation is emphasized in this chapter. If your student first makes a quick estimate she will more likely place the decimal point correctly. This will become more significant in Chapter 9 when she will multiply a decimal by a decimal and the decimal points will not align.

(1) Multiply a decimal by a 2-digit whole number I

Activity

Write the problem 7.8 x 23. Ask your student to estimate the answer first.

Remind your student that 23 is 20 and 3. Have her apply the distributive property and multiply 7.8 by 20 and by 3 and then add the partial products. To multiply by 20, she can multiply by 2 and then by 10. Use the multiplication algorithm to multiply 7.8 by 2 and 3. Thus 7.8 is multiplied by the digits 2 and 3, but the 2 is actually tens.

Guide your student in working the problem vertically, multiplying first the ones and then the tens. Emphasize place values, but point out that it can be confusing where to put the 6 when multiplying 8 tenths by 2 tens in the second row of the partial products since there are no tenths in the answer. So we can put a 0 in the tenths place before multiplying 7.8 by 2 to remember that we are actually multiplying by 20.

Then show the process for 78 x 23. Point out that we work the problem in exactly the same way. So to solve 7.8 x 23, we can leave out the decimals in the partial products for 7.8 x 3 and 7.8 x 20 and work the problem as if we were multiplying whole numbers. In copying the numbers to rewrite the problem vertically, we should still include the decimal in the 7.8 to remember that it is tenths, not a whole number, and that we need to put a decimal point in the answer. Tell your student that a quick estimate, whether done before or after multiplying, is a good way to check that we remembered to put the decimal in the answer as well as check that the digits were aligned correctly.

Repeat with 0.45 x 26. Although customarily we drop trailing 0's, so that 0.45 x 20 = 9 and 0.45 x 6 = 2.7, when we work the problem vertically, the same as we would 45 x 26, those 0's are included in the partial product, and the answer is hundredths compared to the answer to 45 x 26.

The final answer with the trailing 0 dropped is 11.7. Again, an estimate is important to check if the answer is reasonable.

$7.8 \times 23 \approx 8 \times 20 = 160$

7.8×23
$= (7.8 \times 20) + (7.8 \times 3)$
$= (7.8 \times 2 \times 10) + (7.8 \times 3)$
$= (15.6 \times 10) + 23.4$
$= 156 + 23.4$
$= 179.4$

$$\begin{array}{r} {}^{1} \\ 7.8 \\ \times\quad 2 \\ \hline 1\,5.6 \end{array} \qquad \begin{array}{r} {}^{2} \\ 7.8 \\ \times\quad 3 \\ \hline 2\,3.4 \end{array}$$

$$\begin{array}{r} 7.8 \\ \times\quad 2\,3 \\ \hline 2\,3.4 \quad \leftarrow 7.8 \times 3 \\ 1\,5\,6.0 \quad \leftarrow 7.8 \times 20 \\ \hline 1\,7\,9.4 \end{array}$$

$$\begin{array}{r} 7\,8 \\ \times\quad 2\,3 \\ \hline 2\,3\,4 \\ 1\,5\,6\,0 \\ \hline 1\,7\,9\,4 \end{array}$$
$78 \times 23 = 1794$
78 tenths $\times 23$
$= 1794$ tenths $= 179.4$

$$\begin{array}{r} 7.8 \\ \times\quad 2\,3 \\ \hline 2\,3\,4 \\ 1\,5\,6\,0 \\ \hline 1\,7\,9.4 \end{array} \qquad \begin{array}{r} 7.8 \\ \times\quad 2\,3 \\ \hline 2\,3\,4 \\ 1\,5\,6 \\ \hline 3\,9\,0 \end{array} \,\times$$

$0.45 \times 26 \approx 0.5 \times 30 = 15$

0.45×26
$= (0.45 \times 20) + (0.45 \times 6)$
$= (0.45 \times 2 \times 10) + (0.45 \times 6)$
$= (0.9 \times 10) + 2.7$
$= 9 + 2.7$
$= 11.7$

$$\begin{array}{r} 0.4\,5 \\ \times\quad 2\,6 \\ \hline 2\,7\,0 \\ 9\,0\,0 \\ \hline 1\,1.7\,0 \end{array}$$

Practice

Task 1, p. 32

If necessary, remind your student that when estimating with whole numbers, he can round so that only the first digit is not zero. To multiply the rounded numbers, simply multiply the non-zero digits together and add the same number of 0's in both factors (this concept was taught in *Primary Mathematics* 4A).

Continue with some other problems, including smaller decimal numbers. You can include some optional ways of rounding if your student is good at mental math, for example rounding to 12 or 15 or 25. Remind her that estimates should be quick and easy, but we don't always have to round to a single non-zero digit.

Workbook

Exercise 12, p. 19 (answers p. 44)

---

1. (a) 90,000
   (b) 9000
   (c) 900

$3\underline{000} \times 3\underline{0} = 9\underline{0,000}$

$3.267 \times 28 \approx 3 \times 30 = 90$

$0.326 \times 28 \approx 0.3 \times 30 = 9$

$0.03 \times 28 \approx 0.03 \times 30 = 0.9$

$45.8 \times 62 \approx \underline{5}0 \times \underline{6}0 = \underline{3}000$

$2.14 \times 47 \approx 2 \times 50 = 100$

$0.58 \times 63 \approx 0.6 \times 60 = 36$

$4.38 \times 25 \approx 4 \times 25 = 100$

$3.671 \times 17 \approx 4 \times 15 = 60$

$0.138 \times 49 \approx 0.12 \times 50 = 6$

(2) Multiply a decimal by a 2-digit whole number II

Discussion

Concept p. 31

You may want to write these two problems on paper or a whiteboard and have your student make the estimate and actually do the multiplication rather than simply looking at the textbook page. Note that in the textbook the 0 is added after multiplying by tens. It is better to put the 0 in first, and get into the habit of doing so, before multiplying by the ten. This will help him to remember to put the digits from multiplying by ten all one place to the left, since the calculations are done as if multiplying by ones.

Tell your student that when rewriting the problem 21.87 x 23 vertically she should always include the decimal point, even though we work the problem the same way as 2187 x 23 and can ignore it in the working. She should check her answer with an estimate.

Task 2, p. 32

Again, you can write the problem down and let your student preform the multiplication himself so you can observe and correct any errors or misconceptions.

```
2. (a) 12
   (b) 13.57
```

Practice

Task 3, p. 32

The tasks and exercises in the workbook are not going to always tell your student when to estimate. Encourage her to make it a habit.

```
3. (a) 33.54    (b) 12.19    (c) 181.3
   (d) 616.2    (e) 1827     (f) 2383.2
   (g) 153.94   (h) 392.34   (i) 113.04
```

Workbook

Exercise 13, pp. 20-21 (answers p. 44)

Reinforcement

Extra Practice, Unit 7, Exercise 7, pp. 157-160

Enrichment

You can give your student some problems involving multiplication by 3 digits or even more.

Test

Tests, Unit 7, 7A and 7B, pp. 29-31

```
3.986 x 572 ≈ 4 x 600 = 2400

        3.9 8 6
     x    5 7 2
        7 9 7 2
      2 7 9 0 2 0
    1 9 9 3 0 0 0
    2 2 7 9.9 9 2
```

Chapter 8 – Division by a 2-Digit Whole Number

Objectives

♦ Divide a decimal by a 2-digit whole number.
♦ Estimate the quotient in division of decimals.

Notes

In *Primary Mathematics* 3A students learned to divide a whole number by a 1-digit whole number. This was reviewed in *Primary Mathematics* 4A. In *Primary Mathematics* 5A they learned to divide a whole number by a 2-digit whole number. In this chapter the skill is extended to division of a decimal by a 2-digit whole number. If your student cannot divide a whole number by a 2-digit whole number using the standard algorithm, you should consider going back to *Primary Mathematics* 5A where it is first taught so that your student thoroughly understands the process involved.

Division of a decimal by 2 digits involves the same steps as division of a whole number. The digit in the greatest place value is divided first and the remainder renamed and combined with the digit in the next place value. The steps are repeated until all digits are divided, and if there is still a remainder they can be carried on as long as wanted, or the quotient rounded. For example, in 56.98 ÷ 52, 5 tens can't be divided by 52, so rename them as 50 ones and add to the 6 ones. 56 ones divided by 52 is 1 with a remainder of 4 so write 1 in the quotient. Rename 4 as 40 tenths, add 9 tenths; 49 tenths still cannot be divided by 52, so write 0 tenths in the quotient. Rename 49 tenths as 490 hundredths, add the 8 hundredths, and divide 498 hundredths by 52. Use estimation; 498 is almost 520 so try 9 hundredths as the quotient. 9 hundredths x 52 is 468 hundredths and the remainder is 30 hundredths. Rename this as 300 thousandths, and so on. If the digits are aligned carefully, it is not necessary to include the decimal point in each step of the working, or to be concerned with the exact place value while working the problem. The decimal can be included in the quotient when tenths are divided. To check that the decimal is placed correctly, do a quick estimate to see that the quotient is reasonable.

$$56.98 \div 52 \approx 55 \div 55 = 1$$

```
              1.0 9 5 ...
      5 2)5 6.9 8
          5 2          ← 1 x 52
          4.9 8
          4.6 8        ← 0.09 x 52
           .3 0 0
           .2 6 0      ← 0.005 x 52
           .0 4 0
```

(1) Divide a decimal by a 2-digit whole number

Activity

Write the problem 2.34 ÷ 45. Ask your student to estimate the quotient and then divide. Discuss the process, using place-value language. Use place-value discs if needed.

Neither 2 ones nor 23 tenths can be divided into 45 equal groups, but 234 hundredths can. If we put 5 hundredths in each group, 225 hundredths are used (0.05 x 45) leaving 9 hundredths (2.34 – 2.25). 9 hundredths cannot be divided into 45 groups, so we rename them as 90 thousandths and put 2 thousandths in each group, using up all the hundredths (0.002 x 45).

Now write the problem 234 ÷ 24 and have your student estimate the quotient and then find the answer using the division algorithm. Point out that the steps are the same as in the previous problem. We do have to rename 9 ones as 90 tenths to get no remainder. But ignoring the position of the decimal, the division process is the same; the only difference is that the digits are in different places.

$$2.34 \div 45 \approx 2 \div 50 = 0.4$$

$$
\begin{array}{r}
0.0\,5\,2 \\
\hline
4\,5\,)\overline{2.3\,4} \\
\end{array}
$$

    2.2 5   ← 0.05 x 45
    .0 9 0 ← 2.34 – 2.25
    .0 9 0 ← 0.002 x 45

$$234 \div 45 \approx 200 \div 50 = 4$$

$$
\begin{array}{r}
5.2 \\
\hline
4\,5\,)\overline{2\,3\,4} \\
\end{array}
$$

    2 2 5   ← 5 x 45
    9.0   ← 234 – 225
    9.0   ← 0.2 x 45

Tell your student that we can think of 2.34 ÷ 45 as 234 hundredths ÷ 45 and solve in the same way as we do when dividing whole numbers. We should include the decimal points when copying the numbers, but we don't need to include them in the working of the problem. When we divide a decimal by a whole number and align the digits in the quotient with the digits in the dividend (the total) the decimal point in the quotient will align with the decimal point in the dividend. An estimate helps to make sure the decimal point is in the correct place and the answer is reasonable.

Discussion

Concept p. 33

You may want to write these two problems on a whiteboard or paper and have your student actually do the divisions rather than simply looking at the text page.

After your student has finished this page, ask him to divide 592.8 and 5.928 by 19, without actually doing the division. The answer will be the same as for 5928 ÷ 19 except for the position of the decimal point in the answers.

$$5928 \div 19 = 312$$
$$592.8 \div 19 = 31.2$$
$$59.28 \div 19 = 3.12$$
$$5.928 \div 19 = 0.312$$

Task 1, p. 34

1. (a) 70
   (b) 7
   (c) 0.7

Remind your student that to estimate the answer to a problem involving division by a 2-digit number, we can first round the 2-digit number to the nearest ten, and then round the dividend, or whole, to a number where the first two digits are a multiple of the digit in the tens place. For 2877 ÷ 42, we first round 42 to 40. Then find the multiple of 4 that is closest to the first 2 digits of 2877, which in this case is 28. So we round 2877 to 2800, not 2900 even though 2900 is 2877 rounded to the closest hundred.

Have your student estimate some other division problems, including ones with smaller decimal numbers.

$3.267 \div 28 \approx 3 \div 30 = 0.1$
$32.67 \div 28 \approx 30 \div 30 = 1$
$326.7 \div 28 \approx 300 \div 30 = 10$
$578.9 \div 93 \approx 540 \div 90 = 6$
$57.89 \div 93 \approx 54 \div 90 = 0.6$
$5.789 \div 93 \approx 5.4 \div 90 = 0.06$
$0.578 \div 93 \approx 0.54 \div 90 = 0.006$
$27.9 \div 47 \approx 30 \div 50 = 0.6$
$45.813 \div 72 \approx 42 \div 70 = 0.6$
$2.14 \div 47 \approx 2 \div 50 = 0.04$
$0.198 \div 33 \approx 0.18 \div 30 = 0.006$

Practice

Tasks 2-3, p. 34

Task 3(i): The answer, rounded to 3 decimal places, is 2.495. Rounded to 2 decimal places, it is 2.50. Tell your student that normally we drop trailing 0's on decimal numbers, but not if we want to indicate the degree of accuracy of a rounded number. Leaving the 0 in 2.50 indicates that the quotient is correct to 2 decimal places. A quotient of 2.50 means the possible answer is equal to or greater than 2.495 but less than 2.505; 2.5 means the possible answer is equal to or greater than 2.45 and less than 2.55.

2. (a) 0.2
   (b) 0.18

3. (a) 1.13   (b) 0.01   (c) 0.12
   (d) 2.02   (e) 0.55   (f) 0.29
   (g) 2.31   (h) 0.57   (i) 2.50

Workbook

Exercise 14, pp. 22-23 (answers p. 44)

Reinforcement

Extra Practice, Unit 7, Exercise 8, pp. 161-162

For more practice in multiplication and division, you can have your student find the exact answers to the estimation problems on pages 19 and 22 of the workbook. Exact answers are given at the right.

Test

Tests, Unit 7, 8A and 8B, pp. 33-38

Workbook p. 19
  1. (a) 1899.36
     (b) 1322.48
     (c) 1346.24
     (d) 1982.15
     (e) 1993.48

Workbook p. 22 (to 2 decimal places)
  1. (a) 502.60
     (b) 50.26
     (c) 5.03
     (d) 0.50
     (e) 22.02
     (f) 41.87
     (g) 1.83
     (h) 6.44
     (i) 0.11

Chapter 9 – Multiplication by a Decimal

Objectives

◆ Multiply a decimal by a decimal.
◆ Estimate the product in multiplication of decimals.

Material

◆ Mental Math 7

Notes

In this chapter your student will learn to multiply a decimal by a decimal. Although ultimately he will learn to multiply decimals by multiplying the related whole numbers and counting decimal places in the numbers being multiplied in order to insert it in the correct place in the product, it is beneficial to understand why this works. If your student understands the reasoning behind the process easily, that is an indication that he has a good foundation and understands all of the concepts that lead up to this process, including multiplication and division, the connections between fractions, decimals, and division, and above all place value. Take the time needed to be sure he thoroughly understands the concepts. This guide necessarily provides a limited number of examples; you can expand on the lesson by adding additional, similar examples, or guiding your student through some of the practice problems.

The process of multiplying decimals is the same as the process for multiplying whole numbers; the only difference is in the place values of the digits. The number of decimal places in the product of 5.68 x 2.5 is the sum of the number of decimal places in the numbers being multiplied relative to the product of 568 x 25 (including any trailing 0's). This can be illustrated by converting the decimals to fractions before multiplying or by writing each decimal as a product of a whole number and a unitary decimal such as tenths, hundredths, or thousandths (or smaller), then multiplying the whole numbers and decimals separately and then together, similar to what students have done earlier when multiplying tens, hundreds, or thousands together.

Estimation can also help with positioning the decimal correctly in the product. In fact an estimate, helps prevent potential issues with trailing 0's. While 14.200 has 3 decimal places, 14.2 does not and if given just 5.68 x 2.5 = 14.2 without the computation it would not be possible to check the placement of the decimal by just counting decimals places. However, in some cases even making an estimate requires counting decimal places when both numbers are smaller than 0, as in 0.056 x 0.025.

$$5.68 \times 2.5 = \frac{568}{100} \times \frac{25}{10}$$
$$= \frac{14200}{1000}$$
$$= 14.200$$
$$= 14.2$$

$$5.68 \times 2.5 = 568 \times 0.01 \times 25 \times 0.1$$
$$= 568 \times 25 \times 0.01 \times 0.1$$
$$= 14200 \times 0.001$$
$$= 14.200$$
$$= 14.2$$

$$5.68 \times 2.5 \approx 5 \times 2 = 10$$

$0.056 \times 0.25 \approx 0.05 \times 0.2 = ?$
$5 \times 2 = 10$
Move the decimal 3 places:
$0.010 = 0.01$
$0.05 \times 0.2 = 0.01$

(1) Multiply a decimal by a "1-digit" decimal

Discussion

Concept p. 35

In discussing the examples on this page, compare the equations in the same columns as well as in the same rows. Discuss the position of the digit 3 in terms of its place value.

Multiplying a number by 0.1 is the same as multiplying it by 1 tenth. Its value becomes one tenth of its original value. So multiplying a digit by 0.1 has the effect of moving the digit one place to the right. This is true even if it is a decimal that is being multiplied by 0.1. 3 tenths multiplied by 1 tenth is 3 hundredths. Similarly, multiplying a digit by 0.01 makes it one hundredth as big, and in effect moves it two places to the right.

| 3 ones | x 0.1 = 3 tenths |
| 3 | x 0.1 = 0.3 |

Ones place   Tenths place

3 tenths x 0.1 = 3 hundredths
0.3   x 0.1 = 0.03

Tenths place   Hundredths place

3 ones   x 0.01 = 3 hundredths
3   x 0.01 = 0.03

Ones place   Hundredths place

3 tenths x 0.01 = 3 thousandths
0.3   x 0.01 = 0.003

Tenths place   Thousandths place

Tasks 1-2, p. 35-36

You can list each digit by itself to discuss what happens to it when multiplied by 0.1 or 0.01, as shown for task 1 on the right.

Moving the digit one place or two places to the right results in the decimal point being moved one place or two places to the left, respectively.

Your student might notice that multiplying a number by 0.1 or 0.01 has the same effect on the decimal point as dividing a number by 10 and 100. This makes sense; multiplying by 1 tenth is the same as dividing by 10. You can remind her that 0.1 and 0.01 are fractions smaller than 1. Multiplying a number by a proper fraction is the same as taking a fraction of the number; the answer will be smaller than the original number.

1. 0.462

2. 0.163

4   x 0.1 = 0.4
0.6   x 0.1 = 0.06
0.02 x 0.1 = 0.002
4.62 x 0.1 = 0.462

4.62 x 0.1 = 0.4 6 2

$$4.62 \times 0.1 = 0.1 \times 4.62 = \frac{1}{10} \text{ of } 4.62$$

$$4.62 \times 0.1 = 4.62 \times \frac{1}{10} = 4.62 \div 10$$

Practice

Task 3, p. 36

Activity

Write the expression 34 x 0.1 and ask your student for the answer. Then rewrite it with the answer first.

Write some other decimals and ask your student to write an equivalent expression where a whole number is multiplied by 0.1, 0.01, or 0.001.

3. (a) 0.06   (b) 0.006   (c) 0.5
    (d) 2.3   (e) 0.34   (f) 0.014
    (g) 0.04   (h) 0.004   (i) 0.08
    (j) 0.42   (k) 0.834   (l) 0.062

34 x 0.1 = 3.4
3.4 = 34 x 0.1

0.34 = 34 x ____   (0.01)
6.08 = 608 x ____   (0.01)
93.2 = 932 x ____   (0.1)
0.025 = 25 x ____   (0.001)

Write the expression 4 x 0.2 and discuss its solution. This is the same as simply multiplying a decimal by a whole number, which your student has already done. Suggest an alternate method. Since 0.2 is 2 x 0.1, we can multiply 4 first by 2 and then by 0.1. The answer is the same.

Then have your student solve the expression 0.4 x 0.2.

Write the problem 3.4 x 0.2. It can be solved by multiplying by 2 and then 0.1.

Then write the solution shown at the right. Your student should observe that multiplying tenths (3.4 and 0.2) together is the same as multiplying 34 x 2 by hundredths. He might find the process easier to understand if you convert the decimals to fractions, then multiply, then convert the product back to a decimal.

Repeat with 3.4 x 0.02. Have your student count the total number of decimal places in the numbers being multiplied together and compare to the number of decimal places in the product.

Repeat with 1.5 x 0.04. Point out that compared to the product of 15 and 4 the decimal point is three places to the left, even though the final answer where the trailing 0 is dropped has only 2 decimal places.

Discussion

Task 4, p. 36

The girl is thinking of 3 moves total for the decimal point in the product, and then 3 moves back for the related whole number problem. You can rewrite the problem as shown at the right to emphasize this.

By now, your student may be able to come up with a general rule for multiplying decimals. We can multiply decimal numbers by initially ignoring the decimal point, and then insert the decimal point in the product such that the total number of decimal places is the same as the sum of the decimal places in the numbers being multiplied (without dropping trailing 0's)

$4 \times 0.2 = 0.2 \times 4 = 0.8$
$4 \times 0.2 = 4 \times 2 \times 0.1 = 8 \times 0.1 = 0.8$

$0.4 \times 0.2 = 0.4 \times 2 \times 0.1 = 0.08$
$0.4 \times 0.2 = 0.4 \times \frac{2}{10} = 0.4 \times 2 \times \frac{1}{10} = 0.08$

$3.4 \times 0.2 = 3.4 \times 2 \times 0.1 = 6.8 \times 0.1 = 0.68$

$3.4 \times 0.2 = 34 \times 0.1 \times 2 \times 0.1$
$= 34 \times 2 \times 0.1 \times 0.1$
$= 68 \times 0.01$
$= 0.68$
$3.4 \times 0.2 = \frac{34}{10} \times \frac{2}{10} = \frac{68}{100} = 0.68$

$3.4 \times 0.02 = 34 \times 0.1 \times 2 \times 0.01$
$= 34 \times 2 \times 0.1 \times 0.01$
$= 68 \times 0.001$
$= 0.068$
$3.4 \times 0.02 = \frac{34}{10} \times \frac{2}{100} = \frac{68}{1000} = 0.068$

$1.5 \times 0.04 = 15 \times 0.1 \times 4 \times 0.01$
$= 15 \times 4 \times 0.1 \times 0.01$
$= 60 \times 0.001$
$= 0.060$
$= 0.06$
$1.5 \times 0.04 = \frac{15}{10} \times \frac{4}{100} = \frac{60}{1000} = 0.060$

4. 1.88

$2.35 \times 0.8 = 235 \times 0.01 \times 8 \times 0.1$
$= 235 \times 8 \times 0.01 \times 0.1$
$= 1880 \times 0.001$
$= 1.880$
$= 1.88$
$2.35 \times 0.8 = \frac{235}{100} \times \frac{8}{10} = \frac{1880}{1000} = 1.880$

Tasks 5-7, pp. 36-37

Task 5: This task reminds the student to use an estimate to help make sure the answer is reasonable.

Task 6: Ask your student to first estimate the product. Some possible estimates are 10 x 0.05 = 0.5, 11 x 0.05 = 0.55, or 12 x 0.05 = 0.6.

5. 15.96

6. 0.59

7. 0.378

Practice

Task 8, p. 37

Some of these problems involve simply multiplying 1 digit by 1 digit, so an estimate is redundant, other than a rough estimate that the answer will also be less than 1.

8. (a) 0.18	(b) 2.8	(c) 0.115
(d) 2.73	(e) 19.28	(f) 2.826
(g) 0.024	(h) 0.35	(i) 0.145
(j) 0.156	(k) 0.964	(l) 0.381

Workbook

Exercise 15, pp. 24-25 (answers p. 44)

Even though the first example in problem 1 of this exercise shows the answer as a fraction, require your student to give all answers as decimals. Do not require her to write fractions or show all multiple steps for problems 2 and 3, unless you feel it is necessary. By the time she gets to this exercise, she should already know how to place the decimal in the product and why.

Reinforcement

Mental Math 7

Enrichment

Have your student multiply 8 x 8, 800 x 8, 800 x 0.8 and 800 x 0.008. If the answer is a whole number, have him write a decimal point after the last digit. Discuss the position of the decimal point relative to the first problem, 8 x 8. For every decimal place the decimal point moves to the right in the 0.8 or 0.008 to get to 8, it moves to the left in the product. For every place the decimal point moves to the left in 800 to get to 8, it moves to the right in the product, relative to the product of 8 x 8.

8 x 8 = 64

800 x 8 = 6400

800 x 0.8 = 8 x 100 x 8 x 0.1
         = 64 x 10
         = 640

800 x 0.008 = 8 x 100 x 8 x 0.001
          = 64 x 0.1
          = 6.4

(2) Multiply a decimal by a "2-digit" decimal

Discussion

Task 9, p. 37

Rewrite the problems and have your student write both the rounded numbers and answer in order to see the pattern. All the rounded numbers are whole numbers except for (d). For that problem he might realize that 0.5 is one half and he just needs to find half of 400. He should already know how many 0's need to be added when the factors are tens, hundreds, or thousands relative to 4 x 5.

Add 0's after the decimal points such that all the estimated answers have the same number of 0's as the answer to the product of the two whole numbers, 4000 x 50. Relate the position of the decimal to the number of decimal places in the original numbers.

Have your student find the exact answer to (a) 3957 x 49 and use that to find the answers to (b) through (e). She can count decimal places to place the decimal point and compare to the estimate.

As found in the previous lesson, we can multiply decimal numbers by initially ignoring the decimal point, and then insert the decimal point in the product such that the total number of decimal places, including any trailing 0's, is the same as the sum of the decimal places in the numbers being multiplied.

9. (a) 200,000   (b) 20,000
   (c) 2000   (d) 200
   (e) 200

$3957 \times 49 \approx 4000. \times 50. = 200000.$

$395.7 \times 49 \approx 400. \times 50. = 20000.0$

$395.7 \times 4.9 \approx 400. \times 5. = 2000.00$

$395.7 \times 0.49 \approx 400. \times 0.5 = 200.000$

$39.57 \times 4.9 \approx 40. \times 5 = 200.000$

```
        3 9 5 7
    x      4 9
      3 5 6 1 3
    1 5 8 2 8 0
    1 9 3 8 9 3
```

3957 x 49 = 193893	200000.
395.7 x 49 = 19389.3	20000.0
395.7 x 4.9 = 1938.93	2000.00
395.7 x 0.49 = 193.893	200.000
39.57 x 4.9 = 193.893	200.000

Practice

Tasks 10-11, p. 37

Task 10(b) reiterates that we can multiply decimals in the same way we multiply whole numbers, except for the place values of the digits.

10. (a) 12
    (b) 11.954

11. (a) 31.82   (b) 44.356   (c) 263.2
    (d) 11.52   (e) 1.743   (f) 27.3

Workbook

Exercise 16, p. 26 (answers p. 44)

Reinforcement

Extra Practice, Unit 7, Exercise 9, pp. 163-164

Test

Tests, Unit 7, 9A and 9B, pp. 39-42

Chapter 10 – Division by a Decimal

Objectives

♦ Divide a decimal by a 2-digit whole number.
♦ Estimate the quotient in division of decimals.

Notes

In *Primary Mathematics* 5A students learned to divide a whole number by a 2-digit whole number. In Chapter 8 of this unit they learned to divide a decimal by a 2-digit whole number. In this chapter the skill is extended to division of a decimal by a decimal.

Division of a decimal by decimal involves the same steps as division of a whole number, with the only difference being the value of each digit. Trying to keep track of place value, though, is more complex than dividing by a whole number since dividing by a decimal smaller than 1 results in a quotient greater than the dividend so the decimal point in the quotient won't align with the decimal point in the dividend when using the standard algorithm.

Since division is related to fractions, we can find an equivalent division problem, or fraction, where the denominator, or divisor, is a whole number. In the example at the right, this involves multiplying the denominator by 100, so the numerator also needs to be multiplied by 100. This results in the decimal point being moved over the same number of places in both the dividend and the divisor.

To divide by a decimal, move the decimal point over in the divisor until the divisor is a whole number. Move the decimal point in the dividend over the same number of places. Then divide. The decimal point in the quotient will now align with the decimal point in the dividend.

Check the position of the decimal point with an estimate of the original problem. As with multiplication by a decimal, even to make an estimate requires the idea of moving the decimal point over in both dividend and divisor, 25 ÷ 0.5 = 250 ÷ 5 = 50. The concept is therefore taught by first relating the division to a fraction.

$26.78 \div 0.52 \approx 25 \div 0.5 = 50$

$$
\begin{array}{r}
5\ 1.5 \\
0.5\ 2\overline{)2\ 6.7\ 8} \\
2\ 6 \qquad \leftarrow 50 \times 0.52 \\
0.7\ 8 \\
0.5\ 2 \qquad \leftarrow 1 \times 0.52 \\
0.2\ 6 \\
0.2\ 6 \qquad \leftarrow 0.5 \times 0.52
\end{array}
$$

$$26.78 \div 0.52 = \frac{26.78}{0.52} = \frac{2678}{52}$$

$$0.5\,2\overline{)2\ 6.7\ 8}$$

$$
\begin{array}{r}
5\ 1.5 \\
5\ 2\overline{)2\ 6\ 7\ 8.0} \\
2\ 6\ 0\ 0 \qquad \leftarrow 50 \times 52 \\
7\ 8 \\
5\ 2 \qquad \leftarrow 1 \times 52 \\
2\ 6.0 \\
2\ 6.0 \qquad \leftarrow 0.5 \times 52
\end{array}
$$

(1) Divide a decimal by a "1-digit" decimal

Activity

Write the expression $8 \div 2$ and have your student solve it. Then write the other expressions at the right and ask him how he solved them. He has learned earlier that he can simplify division problems such as these by crossing out the same number of 0's in both numbers. Ask him why this works. Review two explanations. One way to explain why we can cross out the 0's is to use place-value discs as shown at the right $800 \div 200$. Another way to explain why it works is to write the division problem as a fraction and simplify.

Now ask your student to find the answers to $16 \div 4$ and $32 \div 8$ and explain why the quotients are the same as that for $8 \div 2$. We can also write each of these as a fraction and show they are equivalent fractions. Multiplying or dividing the numerator or denominator by the same number simply gives an equivalent fraction and does not change its value. Multiplying or dividing both the dividend and divisor by the same number also does not change the value of the expression so the quotient stays the same.

$8 \div 2 = 4$
$80 \div 20 = 4$
$800 \div 200 = 4$
$8000 \div 2000 = 4$
8 hundreds ÷ 2 hundreds = 4
(Make groups of 2 hundreds)

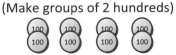

$$800 \div 200 = \frac{800}{200} = \frac{8}{2}$$

$8 \div 2 = 4$
$16 \div 4 = 4$
$32 \div 8 = 4$

$$\frac{32}{8} = \frac{16}{4} = \frac{8}{2} = \frac{4}{1} = 4$$

Discussion

Concept p. 38

You may want to write the problems on a whiteboard or paper and discuss each one separately. If both the numerator and denominator (dividend and divisor) are multiplied by 10 (or 100 or 1000 etc.), the place values of the digits change so that the decimal point is moved the same number of places in both. The goal is to get an equivalent fraction where there is a whole number in the denominator or dividend. The quotient stays the same, but we can now divide by a whole number, which in these examples is always 1.

Have your student solve the problems at the right. Point out that we just need to move the decimal point in both numbers in order to get the denominator (divisor) to a whole number. The numerator (dividend) does not need to be a whole number, since we already know how to divide a decimal by a whole number. You can also discuss whether the answers make sense. It does make sense that there are 30 thousandths in 3 hundredths ($0.03 \div 0.001 = 30$).

$0.03 \div 0.1 = 0.3 \div 1 = 0.3$
$0.03 \div 0.01 = 3 \div 1 = 3$
$0.03 \div 0.001 = 30 \div 1 = 30$
$0.003 \div 0.1 = 0.03 \div 1 = 0.03$
$0.003 \div 0.01 = 0.3 \div 1 = 0.3$
$0.003 \div 0.001 = 3 \div 1 = 3$

Practice

Task 1, p. 39

1. (a) 5000   (b) 50,000   (c) 500,000
   (d) 500   (e) 5000   (f) 50,000
   (g) 50   (h) 500   (i) 5000
   (j) 5   (k) 50   (l) 500
   (m) 0.5   (n) 5   (o) 50
   (p) 0.05   (q) 0.5   (r) 5

Discussion

Task 2, p. 39

Rewrite the problems and have your student write both the rounded numbers and quotient in order to see the pattern. (45 should be an easy enough multiple of 3 for mental computation and is closer to the first 2 digits of the original number than 30 or 6.) In each case, if the number we are dividing by is a decimal, we multiply both numbers by 10 or 100 so that the number we are dividing by is a whole number, 3.

Have your student solve the first division problem and use it to solve the rest, rewriting each with $\overline{)}$ and moving the decimal points. She can then compare the answer to the estimate.

Practice

Tasks 3-4, p. 39

Workbook

Exercise 17, pp. 27-28 (answers p. 44)

Reinforcement

Mental Math 8

2. (a) 1500　　(b) 15,000
   (c) 150　　(d) 150

$4500. \div 3. = 1500.$

$4500 \div 0.3 \approx 45000. \div 3. = 15000.$

$45.00 \div 0.3 \approx 450. \div 3. = 150.$

$4.500 \div 0.03 \approx 450. \div 3. = 150.$

3. (a) 8
   (b) 8.49

4. (a) 153　(b) 2249.5　(c) 12,500

(2) Divide a decimal by a "2-digit" decimal

Discussion

Task 5, p. 40

Rewrite the problems and have your student write both the rounded numbers and answer.

Have your student solve the first division problem and use it to solve the rest, rewriting each with $\overline{)}$ and moving the decimal points. He can then compare the answer to the estimate.

We still want to divide by a whole number. In this case, the whole number will be a 2-digit whole number. So for 4.2 we have to move the decimal point over 1 place in both the numerator (dividend) and denominator (divisor), that is, multiply both by 10. For 0.42 and 0.042 we have to move it over 2 and 3 places, respectively.

5. (a) 90   (b) 900
   (c) 9000   (d) 9
   (e) 900

$3687. \div 42 \quad \approx 3600. \div 40. = 90.$

$3687. \div 4.2 \quad \approx 3600. \div 4. = 900.$

$3687. \div 0.42 \quad \approx 3600. \div 0.4$
$ \approx 36000. \div 4. = 9000.$

$36.87 \div 4.2 \quad \approx 36.00 \div 4 \approx 9$

$36.87 \div 0.042 \approx 36.00 \div 0.04$
$ \approx 3600. \div 4. = 900.$

```
        8 7.7 8 5 7...
 4 2)3 6 8 7
     3 3 6
     ─────
       3 2 7
       2 9 4
       ─────
         3 3 0
         2 9 4
         ─────
           3 6 0
           3 3 6
           ─────
             2 4 0
             2 1 0
             ─────
               3 0 0
               2 9 4
```

```
        8 7 7.8 5 7...
 4.2)3 6 8 7.0
```

```
        8 7 7 8.5 7...
 0.4 2)3 6 8 7.0 0
```

```
        8.7 7 8...
 4.2)3 6.8 7
```

```
           8 7 7.8 5 7...
 0.0 4 2)3 6.8 7 0
```

Practice

Tasks 6-7, p. 40

Workbook

Exercise 18, p. 29 (answers p. 44)

6. (a) 600
   (b) 536.25

7. (a) 25.5   (b) 268.63   (c) 2500

Reinforcement

For more practice in multiplication and division, you can have your student find the exact answers to the estimation problems on pages 26 and 28 of the workbook. Exact answers are given at the right.

Workbook p. 26
1. (a) 14.184   (b) 17.352
   (c) 98.703   (d) 1870.44

Workbook p. 28
2. (a) 274.45   (b) 88.88
   (c) 29.675   (d) 123.166...

Practice

Practice B, p. 41

Problem 9: These can be difficult. Remind your student of the following.

factor x factor = product
product ÷ factor = factor

So for problems missing a factor, i.e., factor x ? = product, we can solve by dividing: product ÷ factor = ?. We can apply the same principle to decimal numbers. For example:

(b) 0.1 x __ = 100
     100 ÷ 0.1 = 1000

Move the decimal point to the right one place in both to get 1000 ÷ 1, or 1000.

1. (a) 57	(b) 150.8	(c) 7250
2. (a) 0.06	(b) 1316	(c) 20,400
3. (a) 10.92	(b) 115.92	(c) 37.41
4. (a) 3.9	(b) 0.342	(c) 0.009
5. (a) 3.3	(b) 1.08	(c) 0.03
6. (a) 10.7	(b) 2.06	(c) 1.25

7. Estimates may vary in precision.
    (a) 0.4 x 40 = 16   (b) 7 x 90 =630 (c) 50 x 40 = 2000
    (d) 6.3 ÷ 70 = 0.09   (e) 30 ÷ 60 = 0.5 (f) 350 ÷ 70 = 5

8. (a) 4.56	(b) 0.039	(c) 0.032
(d) 9200	(e) 3600	(f) 3.2
9. (a) 0.1	(b) 1000	
(c) 0.01	(d) 0.0001	
(e) 0.1	(f) 0.0001	
10. (a) 4.7	(b) 1.7	(c) 128.6
(d) 15.3	(e) 2.0	(f) 3166.7

For problems missing the dividend, e.g., ? ÷ factor = factor, we can solve by multiplying: factor x factor = ?. We can apply the same principle to decimal numbers. For example:

(e) 10 = __ ÷ 0.01
     10 x 0.01 = 0.1

For problems missing the divisor, i.e., product ÷ ? = factor, we can solve by dividing: product ÷ factor = ?. We can apply the same principle to decimal numbers. For example:

(d) 0.01 ÷ ? = 100
     0.01 ÷ 100 = 0.0001

Move the decimal point two places to the *left* in both to get the equivalent fraction 0.0001 ÷ 1, or 0.0001.

Alternatively, your student can try different possibilities to see which way the decimal moves relative to the given answer.

Reinforcement

Extra Practice, Unit 7, Exercise 10, pp. 165-166

Test

Tests, Unit 7, 10A and 10B, pp. 43-46

Workbook

Exercise 12, p. 19

1. (a) $39.57 \times 48 \approx 40 \times 50 = \mathbf{2000}$
   (b) $21.68 \times 61 \approx 20 \times 60 = \mathbf{1200}$
   (c) $42.07 \times 32 \approx 40 \times 30 = \mathbf{1200}$
   (d) $68.35 \times 29 \approx 70 \times 30 = \mathbf{2100}$
   (e) $52.46 \times 38 \approx 50 \times 40 = \mathbf{2000}$

Exercise 13, pp. 20-21

1. (a) 110.4   (b) 240.87
   (c) 1246.44   (d) 31,761
   (e) 50.74   (f) 105.06
   (g) 1498.77   (h) 4834.05

2. 21.6   25.16   73.37
   3122.2   48.76   52.78
   46.5   354.72   1514.88
   watch

Exercise 14, p. 22

1. (a) $5026 \div 10 \approx 5000 \div 10 = \mathbf{500}$
   (b) $502.6 \div 10 \approx 500 \div 10 = \mathbf{50}$
   (c) $50.26 \div 10 \approx 50 \div 10 = \mathbf{5}$
   (d) $5.026 \div 10 \approx 5 \div 10 = \mathbf{0.5}$
   (e) $308.26 \div 14 \approx 300 \div 10 = \mathbf{30}$
       Or: $308.26 \div 14 \approx 300 \div 15 = 20$
   (f) $711.85 \div 17 \approx 800 \div 20 = \mathbf{40}$
       Or: $711.85 \div 17 \approx 700 \div 20 = 35$
   (g) $53.08 \div 29 \approx 60 \div 30 = \mathbf{2}$
   (h) $83.66 \div 13 \approx 80 \div 10 = \mathbf{8}$
       Or: $83.66 \div 13 \approx 90 \div 15 = 6$
   (i) $2.999 \div 28 \approx 3 \div 30 = \mathbf{0.1}$

2. (a) 0.36   (b) 1.00
   (c) 2.16   (d) 3.51
   (e) 0.08   (f) 0.29

Exercise 15, pp. 24-25

1. (a) $\frac{7}{100} = \mathbf{0.07}$ (b) 0.002

   (c) 0.005   (d) 0.9

2. (a) 1.8
   (b) 0.69
   (c) 5.73
   (d) 0.0011
   (e) 0.425

3. (a) $\mathbf{3.56} \times 0.1 = \mathbf{0.356}$
   (b) 1.77
   (c) 6.041

4. Estimates can vary.
   (a) 2.8   (b) $5 \times 0.5 = \mathbf{2.5}$
       2.73   2.325
   (c) $60 \times 0.3 = \mathbf{18}$ (d) $0.9 \times 0.6 = \mathbf{0.54}$
       16.56   0.522
   (e) $2 \times 0.02 = \mathbf{0.04}$ (f) $3 \times 0.9 = \mathbf{2.7}$
       0.034   2.259
   (g) $80 \times 0.06 = \mathbf{4.8}$ (h) $0.2 \times 0.04 = \mathbf{0.008}$
       4.698   0.0084

Exercise 16, p. 26

1. (a) $19.7 \times 0.72 \approx 20 \times 0.7 = \mathbf{14}$
   (b) $38.56 \times 0.45 \approx 40 \times 0.5 = \mathbf{20}$
   (c) $99.7 \times 0.99 \approx 100 \times 1 = \mathbf{100}$
   (d) $214.5 \times 8.72 \approx 200 \times 9 = \mathbf{1800}$

2. (a) 40.068   (b) 82.099
   (c) 190.4   (d) 0.87
   (e) 3.007   (f) 21.83

Exercise 17, pp. 27-28

1. (a) 8000   (b) 900
   (c) 70   (d) 6
   (e) 3000   (f) 20
   (g) 4000   (h) 80
   (i) 9   (j) 7

2. (a) $54.89 \div 0.2 \approx 50 \div 0.2 = 500 \div 2 = \mathbf{250}$
   (b) $44.44 \div 0.5 \approx 45 \div 0.5 = 450 \div 5 = \mathbf{90}$
   (c) $1.187 \div 0.04 \approx 1.200 \div 0.04 = 120 \div 4 = \mathbf{30}$
   (d) $7.39 \div 0.06 \approx 7.20 \div 0.06 = 720 \div 6 = \mathbf{120}$

3. (a) 283.5   (b) 1094.17
   (c) 81,250   (d) 28,571.43

Exercise 18, p. 29

1. (a) $7690 \div 11.3 \approx 7700 \div 11 = \mathbf{700}$
   (b) $251 \div 0.53 \approx 250 \div 0.5 = 2500 \div 5 = \mathbf{500}$
   (c) $369.2 \div 0.48 \approx 350 \div 0.5 = 3500 \div 5 = \mathbf{700}$
   (d) $41.82 \div 0.065 \approx 42 \div 0.07 = 4200 \div 7 = \mathbf{600}$
   (e) $80.96 \div 0.086 \approx 81 \div 0.09 = 8100 \div 9 = \mathbf{900}$
   (f) $1.01 \div 0.010 = 101 \div 1 = \mathbf{101}$

2. (a) 56.58   (b) 191.46
   (c) 88.75   (d) 4857.14

Review 7

Review

Review 7, pp. 42-43

Reviews are cumulative and cover earlier levels.

Problem 4: Your student should apply the distributive property, not perform any computations.

Problem 5(a): Your student should be able to do this without converting any fractions to decimals. Both fractions are less than 1, and the second has a larger denominator.

Problem 16: You may want to save this problem until after the next unit.

Workbook

Review 7, pp. 30-33 (answers p. 46)

Tests

Tests, Units 1-7, Cumulative Tests A and B, pp. 47-52

1. **24,600,000**

2. (a) 0.08   (b) 0.002
   (c) 0.3   (d) 9

3. (a) 9   (b) 8   (c) 1   (d) 110

4. (a) 3.4   (b) 397

5. (a) 1.09, 1.03, $\frac{1}{6}$, $\frac{1}{16}$ (b) 3.22, 3.202, 3.2, 3.022

6. (a) $\frac{3}{5}$ (b) $4\frac{2}{5}$ (c) $6\frac{21}{200}$ (d) $7\frac{9}{40}$

7. (a) 4.7   (b) 6.09   (c) 8.48   (d) 2.18

8. (a) 10   (b) 100
   (c) 0.001   (d) 0.01

9. (a) $2 \times 3 \times 5^2$ (b) $2^5 \times 3$

10. (a) 2.613   (b) 45.904   (c) 27.2802
    (d) 0.09   (e) 20.82   (f) 5.5

11. (a) $\dfrac{\cancel{5}^{\,1}}{8} \times \dfrac{7}{\cancel{10}_{\,2}} = \dfrac{7}{16}$ (b) $\dfrac{5}{8} \div 2 = \dfrac{5}{8} \times \dfrac{1}{2} = \dfrac{5}{16}$

    (c) $\dfrac{8}{5} \div \dfrac{4}{3} = \dfrac{\cancel{8}^{\,2}}{5} \times \dfrac{3}{\cancel{4}_{\,1}} = \dfrac{6}{5} = 1\dfrac{1}{5}$

12. (a) A = 9.2 cm x 4 cm = **36.8 cm²**

    (b) A = (8 cm x 10 cm) − $\dfrac{1}{2}$ x 10 cm x 3 cm
        = 80 cm² − 15 cm²
        = **65 cm²**

13. $\dfrac{48}{112} = \dfrac{6}{14}$

    $\dfrac{6}{14}$ of the members are women.

14. Red marbles: 36 − 8 = 28
    28 : 8 = **7 : 2**
    The ratio of red marbles to blue marbles is 7 : 2.

15. $210 + (10 x $31.25) = $210 + $312.50 = **$522.50**
    The cost is $522.50.

16. 2.5 kg − 0.325 kg − 1.45 kg = **0.725 kg**
    She had 0.725 kg of sugar left.

17. (10 x $0.35) + (8 x $0.70) = $3.50 + $5.60 = **$9.10**
    She paid $9.10.

18. $1\ell - \dfrac{1}{4}\,\ell = \dfrac{3}{4}\,\ell$ $\dfrac{3}{4}\,\ell \div 6 = \dfrac{\cancel{3}^{\,1}}{4} \times \dfrac{1}{\cancel{6}_{\,2}} = \dfrac{1}{8}\,\ell$

    There was $\dfrac{1}{8}\,\ell$ of juice in each cup.

Workbook

Review 7, pp. 30-33

1. (a) Seven hundred thousand, two hundred forty-eight

   (b) Two million, one hundred nine thousand, thirty-five

2. (a) 366   (b) 0.537

3. (a) 1, 2, 4, 5, 10, 20, 25, 50, 100

   (b) 45 cannot be divided without remainder by 4 (it is odd).

   144 cannot be divided without remainder by 5 or 45.

   Both can be divided by 9: 45 ÷ 9 = 5; 144 ÷ 9 = 16

   9 is a common factor of 45 and 144.

4. (a) 800,000   (b) 0.105   (c) 10, 100

5. (a) <   (b) >   (c) =   (d) <

6. (a) 4.386, 4.638, 4.683, 4.9

   (b) 9.392, 9.923, 9.932, 10

   (c) $3.05, 3\frac{1}{3}, 3.5, 3\frac{3}{4}$

7. (a) $\frac{5}{8} = \frac{5 \times 125}{8 \times 125} = \frac{625}{1000} = \mathbf{0.625}$

   (b) $\frac{37}{4} = 9\frac{1}{4} = \mathbf{9.25}$

   (c) $4\frac{3}{5} = 4\frac{6}{10} = \mathbf{4.6}$

8. (a) $0.55 = \frac{55}{100} = \frac{\mathbf{11}}{\mathbf{20}}$

   (b) $5.56 = 5\frac{56}{100} = \mathbf{5}\frac{\mathbf{14}}{\mathbf{25}}$

   (c) $0.095 = \frac{95}{1000} = \frac{\mathbf{19}}{\mathbf{200}}$

   (d) $9.008 = 9\frac{8}{1000} = \mathbf{9}\frac{\mathbf{1}}{\mathbf{125}}$

9. (a) 40.74   (b) 9.36   (c) 26.01

   (d) 2.26   (e) 32.09   (f) 7.37

10. (a) $\frac{3}{4} \times 15 = \frac{45}{4} = 11\frac{1}{4}$ (b) $\frac{\cancel{2}^1}{\cancel{5}^1} \times \frac{\cancel{5}^1}{\cancel{6}_3} = \frac{1}{3}$

    (c) $\frac{8}{9} - \frac{3}{4} = \frac{32}{36} - \frac{27}{36} = \frac{\mathbf{5}}{\mathbf{36}}$ (d) $\frac{6}{7} \div 4 = \frac{\cancel{6}^3}{7} \times \frac{1}{\cancel{4}_2} = \frac{\mathbf{3}}{\mathbf{14}}$

    (e) $1\frac{4}{7} + 6\frac{1}{3} = 7\frac{12}{21} + \frac{7}{21} = 7\frac{\mathbf{19}}{\mathbf{21}}$

    (f) $\frac{4}{5} \div \frac{3}{6} = \frac{4}{5} \div \frac{1}{2} = \frac{4}{5} \times 2 = \frac{8}{5} = 1\frac{\mathbf{3}}{\mathbf{5}}$

11. 5 ℓ ÷ 8 = 0.625 ℓ = **0.63 ℓ** (to 2 decimal places)

12. Adults: 425 + 480 = 905

    Children: 1024 − 905 = 119

    Difference: 905 − 119 = **786**

13. $\frac{2}{5} \times 35 = 2 \times 7 = \mathbf{14}$

14. Use 10 cm side as base, height is 9 cm.

    Area: $\frac{1}{2} \times 10$ cm x 9 cm = 5 cm x 9 cm = **45 cm²**

15.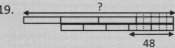

    Let the length of C be 1 unit.

    3 units = 300 cm − 30 cm − (2 x 60 cm) = 150 cm

    1 unit = 150 cm ÷ 3 = **50 cm**

    String C is 50 cm long.

16. 6582 lb ÷ 12 lb = 548 R 6

    He had **548** bags with **6 lb** left over.

17. $\frac{3}{4}$ lb + ($\frac{3}{4}$ lb − $\frac{1}{6}$ lb)

    = [$\frac{9}{12}$ + ($\frac{9}{12}$ − $\frac{2}{12}$)] lb = [$\frac{9}{12}$ + $\frac{7}{12}$] lb = **1$\frac{4}{12}$ lb**

18.

    200 ml

    Let 1 sixth = 1 unit.

    1 unit = 200 ml

    6 units = 200 ml x 6 = **1200 ml**

19.

    48

    Divide fourths into fifths to have 20 units.

    6 units = 48

    1 unit = 48 ÷ 6 = 8

    20 units = 8 x 20 = **160**

    Or: fraction of beads that are blue = $\frac{2}{5} \times \frac{3}{4} = \frac{3}{10}$

    $\frac{3}{10}$ of the beads = 48

    $\frac{1}{10}$ of the beads = 48 ÷ 3 = 16

    All of the beads = 16 x 10 = 160

20. 56 in.² ÷ 7 in. = **8 in.**

21. 0.75 miles x 12 = **9 miles**

Unit 8 – Measures and Volume

Chapter 1 – Conversion of Measures

Objectives

♦ Convert a decimal measurement to a smaller unit or a compound unit.
♦ Convert a measurement to a larger unit as a decimal.

Material

♦ Rulers
♦ Base-10 blocks, optional
♦ Mental Math 9 (appendix)

Vocabulary

♦ Conversion factor

Notes

In *Primary Mathematics 5A* students learned to convert measurements involving fractions to smaller units or compound units. For example, $1\frac{5}{6}$ of a year = 1 year 10 months or 22 months. In this chapter your student will learn to convert measurements involving decimals rather than fractions.

Your student needs to be familiar with the **conversion factors** — how many smaller units equal a larger unit (e.g., 12 in. = 1 ft).

To convert a measurement to a smaller unit, whether given as a whole number or as a decimal, each larger unit is divided into smaller units, so there are more of them. We multiply the larger unit by a conversion factor, which is how many smaller units would be in one of them. Some examples are given at the right.

> 1 kg = 1000 g
> 2 kg = 2 x 1000 g = 2000 g
> 0.2 kg = 0.2 x 1000 g = 200 g
>
> 1 ft = 12 in.
> 3 ft = 3 x 12 in. = 36 in.
> 0.5 ft = 0.5 x 12 in. = 6 in.

In this chapter the only conversion factors that will be used for the metric system are 100 (meter to centimeter) or 1000 (kilogram to gram, kilometer to meter, liter to milliliter). For these it is easy to convert by simply moving the decimal point over 2 or 3 places. The main possibility for error will be forgetting that 1 m is 100 cm, not 1000 cm.

U.S. customary measurements are much less frequently expressed as decimals than as fractions. However, the concepts in converting U.S. customary measurements expressed as decimals to smaller units are the same as with metric measurements; multiply by the conversion factor to go from larger to smaller units of measurement. The answers in this chapter will usually be whole numbers. So for U.S. customary measurements, the decimals will usually be 0.25, 0.5, or 0.75. Since it is easier to find $\frac{1}{4}$ of 12 or even $\frac{3}{4}$ of 12 than it is to multiply 0.25 by 12 or 0.75 by 12, your student can rename the decimal to a fraction first. This will give her more experience and practice in going back and forth between fractions and decimals for easy fraction equivalents.

To convert a measurement from a smaller unit to a larger unit, the smaller units are grouped into a larger unit, so there are fewer larger units. We divide by the conversion factor. In this curriculum, the division will be represented as a fraction. This will make it easier to relate conversion of measurement to rates as taught in Unit 11 later.

In the metric system, division will be by a hundred or a thousand, so the conversion will be easy to do mentally by simply moving the decimal point two or three places to the left.

For U.S. customary measurements, most of the divisions in this chapter will give fractions that simplify to $\frac{1}{4}$, $\frac{1}{2}$, or $\frac{3}{4}$ which are easy to convert to a decimal. Your student can simplify the fraction first and find the decimal mentally.

$$1\,g = \frac{1}{1000}\ kg$$

$$2\,g = 2 \times \frac{1}{1000}\ kg$$

$$= \frac{2}{1000}\ kg$$

$$= 0.002\ kg$$

$$1\ in. = \frac{1}{12}\ ft$$

$$9\ in. = 9 \times \frac{1}{12}\ ft$$

$$= \frac{9}{12}\ ft$$

$$= \frac{3}{4}\ ft$$

$$= 0.75\ ft$$

(1) Convert to smaller units of measurement

Activity

Show your student a ruler or draw a bar and label it 1 ft. Ask him for the number of inches in 1 ft and then in 3 ft and ask him how he found the answer. He had to multiply 3 by 12, the number of inches in a foot. 12 is the *conversion factor* for converting feet to inches.

Then ask your student to find the number of inches in 0.5 ft and ask her how she found the answer. She may have simply realized it was half of a foot, which is 6 inches. Point out that we can find the answer in the same way we did for 3 ft; by multiplying 0.5 by 12, the number of inches in a foot. We can also first rename 0.5 ft as $\frac{1}{2}$ ft and multiply the fraction by the number of inches in a foot.

Now ask your student to find the number of inches in 0.75 ft. He can either multiply 0.75 by 12 or $\frac{3}{4}$ by 12. It is easier to convert to the fraction first and then think of what a fourth of 12 is and multiply that by 3 than it is to calculate 0.75 x 12.

Finally, ask your student to convert 1.75 ft into feet and inches. All we need to convert is the decimal part of the number.

1 ft = 12 in.
3 ft = 3 x 12 in. = 36 in.
0.5 ft = 0.5 x 12 in. = 6 in.
0.5 ft = $\frac{1}{2}$ x 12 in. = 6 in.
0.75 ft = 0.75 x 12 in. = 9 in.
0.75 ft = $\frac{3}{4}$ x 12 in. = 9 in.
1.75 ft
$\diagup$ $\diagdown$
1 ft 0.75 ft = 9 in.

Discussion

Concept p. 44

Discuss this page. Since there are 100 cm in a meter, a tenth of a meter is a tenth of 100 cm, and a hundredth of a meter is a hundredth of 100 cm. Be sure your student understands that we can find the number of centimeters in 0.1 m and 0.01 m by multiplying the number of meters by the number of centimeters in a meter, 100. So to find Ryan's height in centimeters, we can multiply his height in meters by 100 cm.

Point out that when we convert a measurement in a larger unit to the same measurement in a smaller unit the numerical value is greater. In 0.1 m to 10 cm, 10 is larger than 0.1. This makes sense because we are cutting up the larger unit into smaller ones, and so there are more of them. You can also discuss how easy it is to convert meters to centimeters in comparison to feet to inches. Since we multiply by 100, all we have to do is move the decimal point over two places.

Ask your student to convert the measurements into compound units, meters and centimeters. For that, we just convert the decimal part.

126 cm
132 cm

1 m = 100 cm
0.1 m = 0.1 x 100 cm
= 10 cm
0.01 m = 0.01 x 100 cm
= 1 cm
1.4 m = 1.4 x 100 cm
= 140 cm
1.26 m = 1.26 x 100 cm
= 126 cm
1.32 m = 1.32 x 100 cm
= 132 cm
1.32 m
$\diagup$ $\diagdown$
1 m 0.32 m = 32 cm
1.32 m = 1 m 32 cm

Practice

Tasks 1-5, p. 45

There are a lot of problems here, but most of them can be done mentally. Do not require your student to always write a multiplication expression. They are shown at the right simply to show the thought process.

After your student has done these problems, discuss her solutions to the ones involving U.S. customary measurements to see whether she multiplied decimals or fractions. For example, rather than multiplying 2.75 qt by 4 in 3(i), it is easier to mentally find 2 x 4 and then 3 fourths of 4 and add.

Workbook

Exercise 1, p. 34
(answers p. 60)

---

1. (a) 75 cm
   (b) 375 cm
   (c) 6 in.

2. (a) 2800 g   (b) 100 oz

3. (a) 0.6 m = 0.6 x 100 cm   (b) 0.49 ℓ = 0.49 x 1000 ml
             = **60** cm   = **490** ml

   (c) 0.615 km = 0.615 x 1000 m   (d) 0.3 kg = 0.3 x 1000 g
               = **615** m   = **300** g

   (e) 1.85 kg = 1.85 x 1000 g   (f) 4.2 ℓ = 4.2 ℓ x 1000 ml
              = **1850** g   = **4200** ml

   (g) 2.75 qt = 2.75 x 4 c   (h) 3.5 lb = 3.5 x 16 oz
             = **11** c   = **56** oz

   Or: $2 \times 4 = 8; \frac{3}{4} \times 4 = 3$ Or: $3 \times 16 = 48; \frac{1}{2} \times 16 = 8$
       8 c + 3 c = 11 c   48 oz + 8 oz = 56 oz

   (i) 3.25 ft = 3.25 x 12 in.   (j) 0.5 gal = 0.5 x 4 qt
           = **39** in.   = **2** qt

   Or: $3 \times 12 = 36, \frac{1}{4} \times 12 = 3$ or: $= \frac{1}{2} \times 4$ qt
      36 in. + 3 in. = 39 in.   = **2** qt

4. 4 ℓ **200** ml

5. (a) 3.45 km = 3 km + (0.45 x 1000) m   (b) 2.06 m = 2 m + (0.06 x 100) cm
               = **3** km **450** m   = **2** m **6** cm

   (c) 4.005 ℓ = 4 ℓ + (0.005 x 1000) ml   (d) 6.432 kg = 6 kg + (0.432 x 1000) g
               = **4** ℓ **5** ml   = **6** kg **432** g

   (e) 4.25 lb = 4 lb + ($\frac{1}{4}$ x 16) oz   (f) 7.5 ft = 7 ft + ($\frac{1}{2}$ x 12) in.
              = **4** lb **4** oz   = **7** ft **6** in.

(2) Convert to larger units of measurement

Activity

Draw a bar and divide it into 10 units. Tell your student that you are going to measure a line with this bar. Draw a line above it 7 units long. Ask him for the length of the line compared to the total length. He may express it as a fraction — the line is $\frac{7}{10}$ of the total bar. Point out that he took the number of smaller units and divided it by the total number smaller units that was equivalent to the whole bar. Tell him that we can also say the length is "0.7 of" the whole bar.

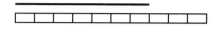

The line is $\frac{7}{10}$ of the bar.

The line is 0.7 of the bar.

Write 70 cm. Ask your student "70 cm is what fraction of a meter?" Then ask her to convert the answer to a decimal. Point out that if we have a measurement in a smaller unit, we can find the measurement in terms of a larger unit by dividing by the number of smaller units in the larger unit, the conversion factor. We can write the division as a fraction, and simplify to change it to a decimal.

$70 \text{ cm} = \frac{70}{100}$ of a meter

$70 \text{ cm} = 0.7$ of a meter.

0.7 of a meter is 0.7 m.

Ask your student to convert 50 m, 500 m, 5000 m, and 5 km 25 m to kilometers and give the answer as a decimal. For 5 km 25 m, he can convert just the 25 m and add to the 5 km, or write the entire amount in meters and divide by 1000. Point out that with decimals, it is easy to convert between units because we can simply move the decimal point back the appropriate number of places.

$50 \text{ m} = \frac{50}{1000} \text{ km} = 0.05 \text{ km}$

$500 \text{ m} = \frac{500}{1000} \text{ km} = 0.5 \text{ km}$

$5000 \text{ m} = \frac{5000}{1000} \text{ km} = 5 \text{ km}$

$5 \text{ km } 25 \text{ m} = 5 \text{ km } \frac{25}{1000} \text{ km} = 5.025 \text{ km}$

Draw a line 9 in. long and ask your student to measure it and give its length in inches, or simply draw a line and label it 9 in. Then ask her, "9 in. is what fraction of a foot?" She should divide 9 in. by the number of inches in a foot. Ask her to express the fraction as a decimal. Point out it is easier to find the decimal if the fraction is first simplified.

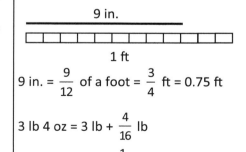

9 in.

1 ft

$9 \text{ in.} = \frac{9}{12}$ of a foot $= \frac{3}{4}$ ft $= 0.75$ ft

Ask your student to find 3 lb 4 oz in pounds only as a decimal.

$3 \text{ lb } 4 \text{ oz} = 3 \text{ lb} + \frac{4}{16} \text{ lb}$

$= 3 \text{ lb} + \frac{1}{4} \text{ lb}$

$= 3.25 \text{ lb}$

Point out that when we go from a smaller measurement unit to a larger one the numerical value is smaller. 0.75 is a smaller number than 9. We are grouping the smaller units into the larger one.

Discussion

Task 8, p. 46

You can help your student find his own height. Repeat with feet and inches and ask him to express his height in feet as a mixed number and as a decimal.

Practice

Tasks 6-7 and 9-12, p. 46

Your student should be able to do most of these problems mentally, particularly the metric conversions.

Workbook

Exercise 2, pp. 35-36 (answers p. 60)

6. 0.145 ℓ

7. (a) $350\,\text{m} = \dfrac{350}{1000}\,\text{km}$
        $= \textbf{0.35}\,\text{km}$

    (b) $420\,\text{ml} = \dfrac{420}{1000}\,\text{ℓ}$
        $= \textbf{0.42}\,\text{ℓ}$

    (c) $625\,\text{g} = \dfrac{625}{1000}\,\text{kg}$
        $= \textbf{0.625}\,\text{kg}$

    (d) $30\,\text{cm} = \dfrac{30}{100}\,\text{m}$
        $= \textbf{0.3}\,\text{m}$

9. **3.5** kg

10. (a) $4\,\text{m}\,35\,\text{cm} = 4\,\text{m} + \dfrac{35}{100}\,\text{m}$
        $= \textbf{4.35}\,\text{m}$

    (b) $5\,\text{km}\,90\,\text{m} = 5\,\text{km} + \dfrac{90}{1000}\,\text{km}$
        $= \textbf{5.09}\,\text{km}$

    (c) $2\,\text{ℓ}\,800\,\text{ml} = 2\,\text{ℓ} + \dfrac{800}{1000}\,\text{ℓ}$
        $= \textbf{2.8}\,\text{ℓ}$

    (d) $4\,\text{kg}\,75\,\text{g} = 4\,\text{kg} + \dfrac{75}{1000}\,\text{kg}$
        $= \textbf{4.075}\,\text{kg}$

    (e) $3\,\text{ft}\,9\,\text{in.} = 3\,\text{ft} + \dfrac{9}{12}\,\text{ft}$
        $= \textbf{3.75}\,\text{ft}$

    (f) $3\,\text{qt}\,2\,\text{c} = 3\,\text{qt} + \dfrac{2}{4}\,\text{qt}$
        $= \textbf{3.5}\,\text{qt}$

11. 3.08 kg

12. (a) $4070\,\text{m} = \dfrac{4070}{1000}\,\text{km}$
        $= \textbf{4.07}\,\text{km}$

    (b) $2380\,\text{ml} = \dfrac{2380}{1000}\,\text{ℓ}$
        $= \textbf{2.38}\,\text{ℓ}$

    (c) $5200\,\text{g} = \dfrac{5200}{1000}\,\text{kg}$
        $= \textbf{5.2}\,\text{kg}$

    (d) $605\,\text{cm} = \dfrac{605}{100}\,\text{m}$
        $= \textbf{6.05}\,\text{m}$

    (e) $51\,\text{in.} = \dfrac{51}{12}\,\text{ft}$
        $= 4\dfrac{3}{12}\,\text{ft}$
        $= \textbf{4.25}\,\text{ft}$

    (f) $88\,\text{oz} = \dfrac{88}{16}\,\text{lb}$
        $= 5\dfrac{8}{16}\,\text{lb}$
        $= \textbf{5.5}\,\text{lb}$

(3) Practice

Practice

Practice A, p. 47

Reinforcement

Extra Practice, Unit 8, Exercise 1, pp. 171-172

Mental Math 9

Test

Tests, Unit 8, 1A and 1B, pp. 53-56

1. (a) $0.285 \ell = 0.285 \times 1000$ ml
$= \mathbf{285}$ ml

(b) 0.75 gal $= \dfrac{3}{4} \times 4$ qt
$= \mathbf{3}$ qt

(c) 0.085 km $= 0.085 \times 1000$ m
$= \mathbf{85}$ m

(d) 0.25 ft $= \dfrac{1}{4} \times 12$ in.
$= \mathbf{3}$ in.

(e) 0.706 kg $= 0.706 \times 1000$ g
$= \mathbf{706}$ g

(f) 0.5 lb $= \dfrac{1}{2} \times 16$ oz
$= \mathbf{8}$ oz

2. (a) 670 ml $= \dfrac{670}{1000} \ell$
$= \mathbf{0.67} \ell$

(b) 12 oz $= \dfrac{12}{16}$ lb
$= \mathbf{0.75}$ lb

(c) 105 m $= \dfrac{105}{1000}$ km
$= \mathbf{0.105}$ km

(d) 3 c $= \dfrac{3}{4}$ qt
$= \mathbf{0.75}$ qt

(e) 69 g $= \dfrac{69}{1000}$ kg
$= \mathbf{0.069}$ kg

(f) 6 in. $= \dfrac{6}{12}$ ft
$= \mathbf{0.5}$ ft

3. (a) 20.08 km $= 20$ km $+ (0.08 \times 1000)$ m
$= \mathbf{20}$ km $\mathbf{80}$ m

(b) 3.75 qt $= 3$ qt $+ (\dfrac{3}{4} \times 4)$ c
$= \mathbf{3}$ qt $\mathbf{3}$ c

(c) $16.5 \ell = 16 \ell + (.5 \times 1000)$ ml
$= \mathbf{16} \ell \, \mathbf{500}$ ml

(d) 18.5 ft $= 18$ ft $+ (\dfrac{1}{2} \times 12)$ in.
$= \mathbf{18}$ ft $\mathbf{6}$ in.

(e) 2.08 kg $= 2$ kg $+ (.08 \times 1000)$ g
$= \mathbf{2}$ kg $\mathbf{80}$ g

(f) 4.75 lb $= 4$ lb $+ (\dfrac{3}{4} \times 16)$ oz
$= \mathbf{4}$ lb $\mathbf{12}$ oz

4. (a) 9 m 60 cm $= 9$ m $+ \dfrac{60}{100}$ m
$= \mathbf{9.6}$ m

(b) 6 gal 3 qt $= (6 \times 4)$ qt $+ 3$ qt
$= \mathbf{27}$ qt

(c) $4 \ell \, 705$ ml $= 4 \ell + \dfrac{705}{1000} \ell$
$= \mathbf{4.705} \ell$

(d) 2 lb 5 oz $= (2 \times 16)$ oz $+ 5$ oz
$= \mathbf{37}$ oz

(e) 25 km 6 m $= 25$ km $+ \dfrac{6}{1000}$ km
$= \mathbf{25.006}$ km

(f) 3 ft 7 in. $= (3 \times 12)$ in. $+ 7$ in.
$= \mathbf{43}$ in.

5. (a) $<$ (950 ml < 1000 ml)
(c) $=$ (9 cm $= 9$ cm)
(e) $>$ (1250 g > 1025 g)
(g) $<$ (82 in. < 84 in.)

(b) $>$ (2038 m > 38 m)
(d) $>$ (3500 ml > 3005 ml)
(f) $=$ (10080 g $= 10080$ g)
(h) $=$ (104 oz $= 104$ oz)

6. 1.64 m $- 6$ cm $= 1.64$ m $- 0.06$ m $= \mathbf{1.58}$ **m**
Her sister is 1.58 m tall.

7. 3.45 kg $- 250$ g $- 1.25$ kg $= 3.45$ kg $- 0.25$ kg $- 1.25$ kg $= \mathbf{1.95}$ **kg**
She had 1.95 kg of flour left.

Chapter 2 – Volume of Rectangular Prisms

Objectives

♦ Find the side of a rectangular prism given its volume and the other two dimensions.
♦ Find one dimension of a rectangular prism given its volume and the area of one face.
♦ Find the side of a cube given its volume.
♦ Solve word problems involving rectangular prisms and the volume of liquids.

Material

♦ Centimeter cubes or base-10 blocks
♦ Multilink cubes
♦ Appendix p. a7

Notes

In *Primary Mathematics* 3B students learned that volume is the amount of space a solid occupies, and is measured in cubic units such as cubic centimeters. They learned how to find the volume of a solid made up of unit cubes drawn in two dimensions. In *Primary Mathematics* 4B students learned to represent standard units of volume with a superscript, e.g., cm^3. This is reviewed briefly in this chapter.

When finding the area of figures made up of cubes, students should assume that the figures drawn on paper can be constructed from blocks that do not link and so there may be hidden cubes that support other cubes. They should also assume that there are hidden blocks *only* when one is necessary to support another block.

In *Primary Mathematics* 4 students also learned to find the volume of a rectangular prism given its length, width, and height. This will also be briefly reviewed.

In this chapter your student will learn to find one of the dimensions if the volume and the other two dimensions, or the area of a face not including the unknown dimension, are given. We divide the volume by the other two dimensions or the area formed by the other two dimensions. For example, if we are given the length and width, the height can be found by dividing the volume by the length and width.

At this level volumes given for cubes will be perfect cubes. For example, a cube with a volume of $27 \ cm^3$ has sides of 3 cm. Your student should recognize perfect cubes for 1-5 and 10. She can either memorize the others as well, use trial and error for 6, 7, 8, or 9, or use prime factorization to find the cube roots.

Students also learned in *Primary Mathematics* 4B that $1 \ cm^3$ is equivalent to 1 milliliter, $1000 \ cm^3$ is equivalent to 1 liter, and to convert cubic centimeters to liters and milliliters and vice versa. In this chapter your student will solve problems involving the volume and height of water in rectangular tanks.

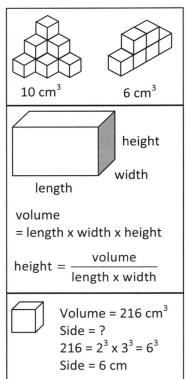

$10 \ cm^3$ $6 \ cm^3$

height
width
length

volume
= length x width x height

$$height = \frac{volume}{length \times width}$$

Volume = $216 \ cm^3$
Side = ?
$216 = 2^3 \times 3^3 = 6^3$
Side = 6 cm

(1) Review volume

Discussion

Concept p. 48

This page is a review of concepts learned earlier. Volume is measured in cubic units. The cubic unit is defined by the length of the side, and standard cubic units have a side of one of the standard measurements, such as centimeters or inches. If you have some centimeter cubes, or the cubes from a set of base-10 blocks, show one to your student. It has a volume of 1 cm^3. Point out that the superscript 3 indicates that there are 3 directions to the measurement, that is, volume is 3-dimensional. Length is one dimension, and can be measured in centimeters, cm, and area is 2-dimensional and can be measured in square centimeters, cm^2. Use a ruler to give your student an idea of the size of a cubic foot, and a meter stick to give your student an idea of the size of a cubic meter, if you have one, or use something else that is about a meter to give a rough idea, such as a baseball bat or the width of a door.

Discussion

Tasks 1-2, pp. 49-50

Task 1: Since all the cubes can't be seen, your student needs to use some calculation to find the number of cubes. Have him tell you how he found the volumes. For A he will need to find the length and width of each layer and multiply to find the number of cubes in each layer, and then add. B and C are rectangular prisms, and once he finds the number of cubes in one layer, he can multiply by the number of layers. D is just a composite figure of two rectangular prisms.

1. A = 40 cm^3
   B = 36 cm^3
   C = 64 cm^3
   D = 60 cm^3
   C has the greatest volume.

2. 24 cm^3

Task 2: This task reviews the formula for finding the area of a rectangular prism, which your student should already have covered in more detail in *Primary Mathematics* 4B. If necessary, you can construct a rectangular prism of several layers with multilink cubes to show that each layer has the same number of cubes so the total is the number of cubes in one layer times the total number of layers. Remind her that the number of cubes in a layer is the same as the area of that surface. Make sure she knows that it does not matter which side is taken to be the length, width, or height, and in what order they are multiplied. We could use the face with 8 cubes and multiply by 3, or the face with 12 cubes and multiply by 2, or the face with 6 cubes and multiply by 4 to find the volume.

Practice

Tasks 3-7, pp. 51-52

Workbook

Exercise 3, p. 37 (answers p. 60)

3. 30 cm^3

4. (a) 240 cm^3 (b) 160 cm^3

5. (a) 125 cm^3 (b) 512 cm^3

6. 30 cm^3

7. (a) 240 cm^3 (b) 280 cm^3

(2) Find an unknown dimension

Activity

Give your student 64 multilink cubes and ask him to build a larger cube from it. If you don't have that many, use 27 or 8 cubes. When he is finished, ask him how he determined how long each side should be. He would have had to find some number for the length of the side such that side x side x side = 64. Tell him that the number 64 is a called a perfect cube. Have him create a table of the perfect cubes for the numbers 1-10.

Ask your student to find the length of the sides of a cube with a volume of 216 cm^3. (Optional: You can show her how to find cube roots using prime factorization. However, at this level the given volumes for cubes will be fairly small so it will be easy to determine the cube roots by just trying some factors.)

Ask your student to build a rectangular prism of volume 36 cubic units, with one side 4 cubes and another 3 cubes. Then discuss how he could determine how long the third side should be without actually using the cubes. The third side is equal to the volume divided by the cubes in one layer, which is the product of the two sides. Point out that when we divide the volume by the product of the two sides, we are actually dividing it by the area of the face formed by those two sides.

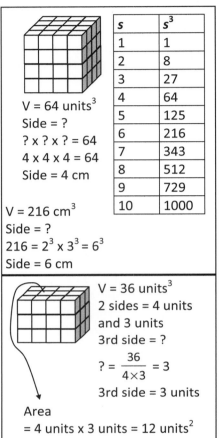

V = 64 units^3
Side = ?
? x ? x ? = 64
4 x 4 x 4 = 64
Side = 4 cm

s	s^3
1	1
2	8
3	27
4	64
5	125
6	216
7	343
8	512
9	729
10	1000

V = 216 cm^3
Side = ?
$216 = 2^3 \times 3^3 = 6^3$
Side = 6 cm

V = 36 units^3
2 sides = 4 units
and 3 units
3rd side = ?

$? = \dfrac{36}{4 \times 3} = 3$

3rd side = 3 units

Area
= 4 units x 3 units = 12 units^2

Discussion

Tasks 8-9, pp. 52-53

Task 9: Be sure your student understands that she can divide the volume by the length of any two sides to get the length of the third side, so she can use the same formula for a missing length or width. Tell her that 3 cm x 2 cm is the area of one of the faces (the top or bottom face). So we could also find the height if we had been given just the area of the top or bottom faces

8. **3 x 3 x 3** = 27
   3 cm

9. 4 cm

Practice

Tasks 10-11, p. 53

If your student finds the answers using the division algorithm, have him try solving the problems by simplifying the fractions to see if that is easier.

10. (a) AB = $\dfrac{576 \text{ cm}^3}{8 \text{ cm} \times 8 \text{ cm}}$ = **9 cm**   (b) CD = $\dfrac{216 \text{ m}^3}{6 \text{ m} \times 3 \text{ m}}$ = **12 cm**

11. (a) EF = $\dfrac{264 \text{ cm}^3}{66 \text{ cm}^2}$ = **4 cm**   (b) CD = $\dfrac{288 \text{ ft}^3}{72 \text{ ft}^2}$ = **4 ft**

$\dfrac{288}{72} = 4$

Workbook

Exercise 4, p. 38 (answers p. 60)

(3) Find heights of liquids

Discussion

Tell your student we have been finding the volume of solids in cubic units and ask her what units are used for volumes of liquids. Two common units are liters and milliliters. Tell her, if she does not mention it, that cubic centimeters, meters, feet, and so on are also used for liquids. Show her a centimeter cube. If a box were made that just fit around the cube, its capacity would be 1 milliliter. 1 milliliter of water fills the same amount of space as 1 cubic centimeter. A milliliter is about 20 drops of water.

If you have base-10 blocks show your student the thousand-cube and ask him what its capacity would be in cubic centimeters. Or show him a liter measuring cup. Since 1 liter = 1000 milliliters, 1 liter = 1000 cm^3.

$$1 \text{ ml} = 1 \text{ cm}^3$$
$$1000 \text{ ml} = 1000 \text{ cm}^3$$
$$1 \text{ } \ell = 1000 \text{ cm}^3$$

Ask your student to convert 15 ℓ, 1.5 ℓ, and 0.15 ℓ to cubic centimeters. Since we are going from a larger measurement unit to a smaller one, essentially cutting it up into smaller parts, we multiply the number of cubic centimeters in a liter.

$$15 \text{ } \ell = 15 \times 1000 \text{ cm}^3 = 15,000 \text{ cm}^3$$
$$1.5 \text{ } \ell = 1.5 \times 1000 \text{ cm}^3 = 1500 \text{ cm}^3$$
$$0.15 \text{ } \ell = 0.15 \times 1000 \text{ cm}^3 = 150 \text{ cm}^3$$

Ask your student to convert 5200 cm^3, 520 cm^3, and 52 cm^3 to liters. Since we are going from a smaller measurement unit to a larger one, essentially making groups of the smaller unit, we divide by the number of cubic centimeters in a liter.

$$5200 \text{ cm}^3 = \frac{5200}{1000} \text{ } \ell = 5.2 \text{ } \ell$$
$$520 \text{ cm}^3 = \frac{520}{1000} \text{ } \ell = 0.52 \text{ } \ell$$
$$52 \text{ cm}^3 = \frac{52}{1000} \text{ } \ell = 0.052 \text{ } \ell$$

The calculations in converting between liters and cubic centimeters involve simply moving the decimal point 3 places in the appropriate direction, similar to converting between milliliters and liters. Remind your student that 1000 cm^3 is not 1 m^3. A cubic meter has 100 cm on the side, and so has a volume of 100 cm x 100 cm x 100 cm = 1,000,000 cm^3.

If you have a rectangular tank fill it part-way with water, or use the pictures in the textbook on p. 54. Tell your student we can find the volume of the water from the length and width of the tank and the height of the water, since the water in the tank is shaped like a rectangular prism.

Practice

Tasks 12-17, pp. 54-55

These tasks apply the concepts previously learned to water in a rectangular tank, and review converting from milliliters or liters to cubic centimeters.

Workbook

Exercise 5, pp. 39-40 (answers p. 60)

12. 600 cm^3

13. 12 m^3

14. 3 cm

15. 1000 cm^3
    1 ℓ = **1000** cm^3
    1 ml = **1** cm^3

16. (a) 2500 cm^3
    (b) 3.2 ℓ

17. 30,000 cm^3
    30 ℓ

(4) Solve word problems involving volume

Discussion

Tasks 18-19, p. 56

Discuss these problems with your student.

For more challenge you can write the problems out on paper or whiteboard and have your student solve them without seeing the hints in the guide or the images. Tell him that when a picture is not present in problems like these, a quick sketch can be helpful in solving the problem. If your student does not know how to draw the tanks, show him that he can draw two congruent rectangles offset slightly and then draw lines joining the corresponding corners. To show water level, draw lines parallel to the lines for the base. Since this type of picture tends to flip which side is to the forefront depending on where the eye focuses, some of the lines can be dashed to help alleviate the optical illusion effect.

18. 12.5 cm

19. 7200 cm^3 = 7200 ml

20. 3.75 cm
    10 cm − **3.75** cm = **6.25** cm

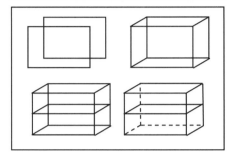

Practice

Practice B, p. 57

Workbook

Exercise 6, pp. 41-42 (answers p. 60)

Enrichment

Have your student look at problem 4 of practice B again and ask her to imagine the size of this tank. It is very large; its height is about three times the height of a man. Ask your student to find the volume of the water in liters.

Write the following problem and have your student solve it.

⇒ Container A and B are both rectangular. Container A is 24 cm long, 10 cm wide, and 40 cm high. It is one quarter filled with water. Container B is 30 cm long and 20 cm wide. It is filled with water to a height of 16 cm. When all the water in Container A is added to the water in Container B, Container B is two thirds full. What is the height of Container B?

1. (a) AB = $\dfrac{360 \text{ cm}^3}{90 \text{ cm}^2}$ = **4 cm**   (b) XY = $\dfrac{576 \text{ cm}^3}{8 \text{ cm} \times 8 \text{ cm}}$ = **9 cm**

2. (a) 25 cm x 8 cm x 18 cm
        = 3600 cm^3 = **3.6 ℓ**
   (b) 22 cm x 15 cm x 10 cm
        = 3300 cm^3 = **3.3 ℓ**

3. 125 = 5^3
   Each edge = **5 cm**

4. $\dfrac{300 \text{ m}^3}{12 \text{ m} \times 5 \text{ m}}$ = **5 m**

300 m^3 = 300 cubes 1 m on the side.
1 cubic meter   = 100 cm x 100 cm x 100 cm
                 = 1,000,000 cm^3
300 cubic meters = 300 x 1,000,000 cm^3
                 = 300,000,000 cm^3
                 = 300,000 ℓ

Height of water in A: $\frac{1}{4}$ x 40 cm = 10 cm

Volume of water in A: 24 cm x 10 cm x 10 cm = 2400 cm^3

Height added to water in B: $\dfrac{2400 \text{ cm}^3}{30 \text{ cm} \times 20 \text{ cm}}$ = 4 cm

New height in B: 16 cm + 4 cm = 20 cm

$\frac{2}{3}$ of height = 20 cm; $\frac{1}{3}$ of height = 20 cm ÷ 2 = 10 cm

Total height = 30 cm
The height of Container B is 30 cm.

Enrichment—Volume of a Solid

Tell your student that there are formulas for finding the volume of spheres, pyramids, triangular prisms, and other regular shapes that he will learn later. But we can also find the volume of irregular shapes.

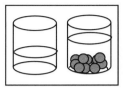

Put some water in a narrow glass or, preferably, a measuring cup with markings on the side for volume. Add some sinkable objects, all of the same shape, sufficient to raise the level of the water, and ask your student to observe the change in height of the liquid. Tell her that when an object is completely submerged in water, it displaces a volume of water that is equal to its own volume. We can therefore use the change in volume of the water to measure the volume of the objects. For example, if the volume of water in the glass is 100 cm^3, and after 10 marbles are added the new volume of marbles and water is a total of 140 cm^3, then the marbles displaced 40 cm^3 of water. Since there are 10 marbles, each marble must have a volume of 4 cm^3.

If the objects are placed in water in a rectangular tank, then we can use the change in height of the water to measure the volume of the objects. Write the following problem and have your student solve it:

⇒ A rectangular tank, 20 cm long and 10 cm wide, is filled with water to a depth of 5 cm. When a rock was put in, the water level rose to 7 cm. What is the volume of the rock?

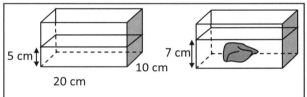

Difference in height: 2 cm
Volume of rock: 20 cm x 10 cm x 2 cm = 400 cm^3

For further practice or enrichment, have your student do appendix p. a7. Answers and possible solutions are given at the bottom of this page.

Reinforcement

Extra Practice, Unit 8, Exercise 2, pp. 173-182 (The last two problems, 9(i) and 9(j) deal with displacement of water by a solid.)

Test

Tests, Unit 8, 2A and 2B, pp. 57-65

Appendix p. a7

1. Increase in height: $\dfrac{600 \text{ cm}^3}{20 \text{ cm} \times 15 \text{ cm}}$ = 2 cm

   New height of water: 20 cm + 2 cm = 22 cm
   The height of the new water level is 22 cm.

2. Change in height: 22 cm − 18 cm = 4 cm
   Volume of stone: 25 cm x 20 cm x 4 cm = 2000 cm^3

3. Volume of cubes: 3 x (20 cm x 20 cm x 20 cm)
   = 24,000 cm^3
   Total volume: 24,000 cm^3 + 136,000 cm^3
   = 160,000 cm^3

   Length of tank: $\dfrac{160,000 \text{ cm}^3}{80 \text{ cm} \times 80 \text{ cm}}$ = 25 cm

4. Volume of 4 marbles is equal to the volume of 3 cubes. A went down same amount B went up.
   Volume of 1 marble: 300 ÷ 2 = 150 cm^3
   Volume of 4 marbles: 150 cm^3 x 4 = 600 cm^3
   Volume of 1 cube: 600 cm^3 ÷ 3 = 200 cm^3

5. Change in height = $\dfrac{5600 \text{ cm}^3}{20 \text{ cm} \times 20 \text{ cm}}$ = 14 cm

   Since the height goes from $\frac{3}{6}$ full to $\frac{5}{6}$ full, then $\frac{2}{6}$ of the height, or $\frac{1}{3}$, is 14 cm.
   Total height: 14 cm × 3 = 42 cm.

Workbook

Exercise 1, p. 34

1. (a) 0.4 km
   = 0.4 x 1000 m
   = **400** m

(b) 1.5 km
   = 1.5 x 1000 m
   = **1500** m

(c) 0.09 kg
   = 0.09 x 1000 g
   = **90** g

(d) 0.43 m
   = 0.43 x 100 cm
   = **43** cm

(e) 1.25 ft
   = 1 ft + ($\frac{1}{4}$ x 12) in.
   = **1** ft **3** in.

(f) 4.75 lb
   = 4 lb + ($\frac{3}{4}$ x 16) oz
   = **4** lb **12** oz

(g) 3.04 km
   = 3 km + (0.04 x 1000) m
   = **3** km **40** m

(h) 3.8 ℓ
   = 3 ℓ + (0.8 x 1000) ml
   = **3** ℓ **800** ml

Exercise 2, pp. 35-36

1. (a) 6 g = $\frac{6}{1000}$ kg
   = **0.006** kg

(b) 8 cm = $\frac{8}{100}$ m
   = **0.08** m

(c) 40 ml = $\frac{40}{1000}$ ℓ
   = **0.04** ℓ

(d) 54 m = $\frac{54}{1000}$ km
   = **0.054** km

(e) 2 kg 300 g
   = 2 kg + $\frac{300}{1000}$ kg
   = **2.3** kg

(f) 3 m 50 cm
   = 3 m + $\frac{50}{100}$ m
   = **3.5** m

(g) 4 km 30 m
   = 4 km + $\frac{30}{1000}$ km
   = **4.03** km

(h) 2 ℓ 600 ml
   = 2 ℓ + $\frac{600}{1000}$ ℓ
   = **2.6** ℓ

2. (a) 250 cm = $\frac{250}{100}$ m
   = **2.5** m

(b) 1080 g = $\frac{1080}{1000}$ kg
   = **1.08** kg

(c) 3006 m = $\frac{3006}{1000}$ km
   = **3.006** km

(d) 2400 g = $\frac{2400}{1000}$ kg
   = **2.4** kg

(e) 14 c = $\frac{14}{4}$ qt
   = **3.5** qt

(f) 345 cm = $\frac{345}{100}$ m
   = **3.45** m

(g) 231 in. = $\frac{231}{12}$ ft
   = **19.25** ft

(h) 3245 ml = $\frac{3245}{1000}$ ℓ
   = **3.245** ℓ

Exercise 3, p. 37

1. (a) 5 cm x 6 cm x 12 cm = **360 cm³**
   (b) 23 cm x 12 cm x 17 cm = **4692 cm³**
   (c) $\frac{1}{2}$ ft x $\frac{3}{4}$ ft x 3 ft = $\frac{9}{8}$ ft³ = **1$\frac{1}{8}$ ft³**
   (d) 30 in² x 7 in. = **210 in.³**

Exercise 4, p. 38

1. ? x ? x ? = 64 cm³   ? = **4 cm**

2. (a) GH = **6 cm**

(b) XY = $\frac{315 \text{ cm}^3}{9 \text{ cm} \times 5 \text{ cm}}$ = **7 cm**

(c) AB = $\frac{96 \text{ cm}^3}{24 \text{ cm}^2}$ = **4 cm**

Exercise 5, pp. 39-40

1. (a) 4500 cm³   (b) 250 cm³
   (c) 80 cm³   (d) 750 cm³

2. (a) 5.6 ℓ   (b) 0.45 ℓ
   (c) 0.02 ℓ   (d) 12 ℓ
   (e) 1.2 ℓ   3.6 ℓ
       3.6 ℓ   1.44 ℓ

3. $\frac{900 \text{ cm}^3}{20 \text{ cm} \times 15 \text{ cm}}$ = **3 cm**

4. $\frac{15 \text{ m}^3}{6 \text{ m}^2}$ = **2.5 m**

Exercise 6, pp. 41-42

1. Decrease in height: $\frac{240 \text{ in.}^3}{12 \text{ in.} \times 10 \text{ in.}}$ = 2 in.
   New height of water: 11 in. – 2 in. = **9 in.**

2. Decrease in height: $\frac{2000 \text{ cm}^3}{40 \text{ cm} \times 25 \text{ cm}}$ = 2 cm
   New height: 15 cm – 2 cm = **13 cm**

3. 33 cm – 18 cm = 15 cm
   600 cm³ x 15 cm = 9000 cm³ = **9 ℓ**
   9 ℓ of water is needed to fill the tank.

4. $\frac{1}{4}$ x 36 cm = 9 cm
   12 cm – 9 cm = 3 cm
   16 cm x 11 cm x 3 cm = **528 cm³**
   528 ml of water needs to be poured off.

Review 8

Review

Review 8, pp. 58-60

Workbook

Review 8, pp. 43-46 (answers p. 63)

Tests

Tests, Units 1-8, Cumulative Tests A and B, pp. 67-73

---

1. (a) 0.005   (b) 50,000   (c) 0.05

2. (a) 50.806   (b) 7.031
   (c) 45.308   (d) 8.009

3. (a) 12.61
   (b) 9.9

4. (a) 31,238   31,328   31,823   31,832
   (b) It is easy to convert all of these to decimals mentally: 4.166..., 4.5, 4.4, 4.3.
   In order: $4\frac{1}{6}, 4\frac{3}{10}, 4\frac{2}{5}, \frac{9}{2}$
   (c) 4.089   4.98   498   4809
   (d) $2\frac{1}{2} = 2.5$ 2.05 $2\frac{3}{5} = 2.6$ 2.51
   In order: **2.05, $2\frac{1}{2}$, 2.51, $2\frac{3}{5}$**

5. (a) 8.4   (b) 5.75   (c) 1.875

6. (a) 0.43   (b) 0.22   (c) 3.83

7. (a) $\dfrac{31}{500}$ (b) $2\dfrac{9}{25}$ (c) $6\dfrac{77}{250}$

8. (a) 1,200,000   (b) 18,120   (c) 201.5
   (d) 24   (e) 0.489   (f) 3.25
   (g) 13.6   (h) 27.9   (i) 0.3906
   (j) 18.5   (k) 3.8   (l) 3.65

9. (a) $23 \times (\underline{34 - 25})$ (b) $\underline{7 \times 8} + \underline{48 \div 3}$
   $= 23 \times 9$ $= 56 + 16$
   $= \mathbf{207}$ $= \mathbf{72}$

   (c) $(\underline{45 - 31}) \times 4 + 12$ (d) $(\underline{28 + 9}) \times (\underline{12 - 7}) =$
   $= \underline{14 \times 4} + 12$ $= 37 \times 5$
   $= 56 + 12$ $= \mathbf{185}$
   $= \mathbf{68}$

10. (a) $\dfrac{2}{\cancel{3}_1} \times \cancel{45}^{15}$ (b) $\dfrac{\cancel{35}^{5}}{\cancel{12}_2} \times \dfrac{\cancel{18}^{3}}{\cancel{7}_1} = \dfrac{15}{2}$ (c) $\dfrac{7}{9} \div 5 = \dfrac{7}{9} \times \dfrac{1}{5}$
    $= \mathbf{30}$ $= 7\dfrac{1}{2}$ $= \dfrac{7}{45}$

    (d) $4 \div \dfrac{3}{8} = 4 \times \dfrac{8}{3}$ (e) $\dfrac{3}{5} \div \dfrac{5}{8} = \dfrac{3}{5} \times \dfrac{8}{5}$ (f) $\dfrac{\cancel{14}^{2}}{3} \times \dfrac{1}{\cancel{7}_1} = \dfrac{2}{3}$
    $= \dfrac{32}{3}$ $= \dfrac{24}{25}$
    $= 10\dfrac{2}{3}$

11. $124 = \mathbf{2^2 \times 31}$

Continued next page

12. 15 m − 2.5 m = 12.5 m

12.5 m ÷ 6 ≈ 2.08 m ≈ **2.1 m**

The length of each piece is about 2.1 m.

13.

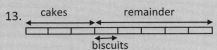

cakes   remainder

biscuits

Each eighth is a unit. She used 4 units, which is $\frac{1}{2}$ of the flour.

Or: $\frac{3}{8} + \left(\frac{1}{5} \times \frac{5}{8} \right) = \frac{3}{8} + \frac{1}{8} = \frac{4}{8} = \frac{1}{2}$

14. To show both $\frac{1}{3}$ and $\frac{1}{4}$ on a bar, use 12 units (lowest common multiple of 3 and 4).

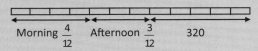

Morning $\frac{4}{12}$ Afternoon $\frac{3}{12}$ 320

5 units remain (12 − 4 − 3 = 5)

5 units = 320

1 unit = 320 ÷ 5 = 64

He had 12 units at first.

12 units = 64 x 12 = **768**

He had 768 eggs at first.

Or: $1 - \frac{1}{3} - \frac{1}{4} = \frac{12}{12} - \frac{4}{12} - \frac{3}{12} = \frac{5}{12}$

$\frac{5}{12}$ of the eggs = 320

$320 \div \frac{5}{12} = 768$

15.

Female

Male

48

?

3 units = 48

1 unit = 48 ÷ 3 = 16

8 units = 16 x 8 = **128**

There are 128 members altogether.

16. Width: 300 ÷ 20 = **15 m**

Perimeter: 2 x (20 m + 15 m) = 2 x 35 m = **70 m**

17. Perimeter:

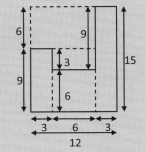

Method 1: Add lengths.

Perimeter = 3 cm + 6 cm + 3 cm + 15 cm + 3 cm + 9 cm + 6 cm + 3 cm + 3 cm + 9 cm = **60 cm**

Method 2: Slide up all the top edges; they would make 12 cm.

Slide 9 cm side all the way to the left. There is 3 + 3 = 6 cm overlap.

Perimeter = 2 x (12 cm + 15 cm) + 6 cm = 60 cm

Area:

Method 1: Divide figure into rectangles.

Area = (9 cm x 3 cm) + (6 cm x 6 cm) + (15 cm x 3 cm)

= 27 cm^2 + 36 cm^2 + 45 cm^2 = **108 cm^2**

Method 2: Create a large rectangle from the largest edges of the figure, and subtract the extra area.

Area = (12 cm x 15 cm) − (6 cm x 9 cm) − (3 cm x 6 cm)

= 180 cm^2 − 54 cm^2 − 18 cm^2 = 108 cm^2

18. A base of the top triangle is 6 cm and the corresponding height is 2 cm.

A base of the bottom triangle is 6 cm and its corresponding height is 6 cm.

Area = $(\frac{1}{2}$ x 6 cm x 2 cm) + $(\frac{1}{2}$ x 6 cm x 6 cm)

= 6 cm^2 + 18 cm^2 = **24 cm^2**

19. (a) 120 in.^{3} (b) 14 ft^3

20. Volume in A: 6 cm x 5 cm x 10 cm = 300 cm^3

Height in B: $\frac{300 \text{ cm}^3}{10 \text{ cm} \times 4 \text{ cm}}$ = **7.5 cm**

The height of the water level in Container B is now 7.5 cm.

Workbook

Review 8, pp. 43-44

1. (a) 860,709
   (b) 3,000,040

2. (a) 7
   (b) 4.21

3. (a) 7000
   (b) 30,000
   (c) 400
   (d) 100

4. (a) 2.7   (b) 3.08
   (c) 1.6   (d) 1.75

5. The numerator should be between the second and third multiples of the denominator. The only one that meets that criteria is $\frac{11}{4}$.

6. $\frac{5}{3}$ is greater than 1 but less than 2; $\frac{5}{8}$ and $\frac{7}{12}$ are both less than 1; $\frac{5}{8} = \frac{15}{24}$; $\frac{7}{12} = \frac{14}{24}$
   The order is
   $2\frac{1}{2}, \frac{5}{3}, \frac{5}{8}, \frac{7}{12}$

7. (a) −   (b) +
   (c) x   (d) ÷
   (e) x   (f) ÷

8. 282.69

9. $\frac{1}{2}$

10. (a) 25 cm   (b) 2400 g

11. (a) 0.58 kg   (b) 4.6 km
    (c) 2.004 ℓ

12. 15,435

13. Cost of shirts: 2 x $12.95 = $25.90
    Cost of T-shirts: 3 x $8.75 = $26.25
    Total cost: $25.90 + $26.25 = **$52.15**

14. $\frac{2}{5}$ m ÷ 4 = $\frac{2}{5}$ m x $\frac{1}{4}$ = $\frac{1}{10}$

15. Remaining cloth: 10 m − 2.35 m = 7.65 m
    Length of each piece: 7.65 m ÷ 5 = **1.53 m**

16. Perimeter: 2 x (5 + 8) = 2 x 13 = 26
    Ratio of length to perimeter: 8 : 26 = **4 : 13**

17. Area = (10 x 6) + ($\frac{1}{2}$ x 8 x 6) = 60 + 24 = **84 cm²**

18.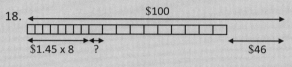

    Cost of 10 towels = $100 − ($1.45 x 8) − $46
    = $100 − $11.60 − $46
    = $42.40
    Cost of 1 towel = $42.40 ÷ 10 = **$4.24**
    1 towel cost $4.24.

19.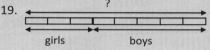

    5 units = 40
    1 unit = 40 ÷ 5 = 8
    8 units = 8 x 8 = **64**
    There were 64 children altogether.

20.

    5 units = 45
    1 unit = 45 ÷ 5 = 9
    4 units = 4 x 9 = **36**
    Jared received 36 more stamps than Juan.

21. $\frac{1080 \text{ in.}^3}{72 \text{ in.}^2}$ = **15 in.**
    Its height in 15 in.

Unit 9 – Percentage

Chapter 1 – Percent

Objectives

- ◆ Read and interpret a percentage of a whole.
- ◆ Express a fraction with a denominator of 100 or 10 as a percentage.
- ◆ Express a percentage as a decimal or a fraction in its simplest form.

Material

- ◆ 10 by 10 grids (appendix p. a8)

Vocabulary

- ◆ Percent
- ◆ Percentage

Notes

In earlier levels of *Primary Mathematics* students learned to express parts of a whole as fractions. In this unit your student will learn to express parts of a whole as a percentages. He will also learn to convert a decimal number to a percentage. Only percentages under 100% will be studied at this level.

The symbol % is read as **percent** and comes from the Latin phrase *per centum*, which means "out of a hundred." The % symbol means "per hundred" or "out of 100." Your student might recognize *cent* as representing 100, from *cent*s in a dollar or *cent*imeters in a meter. Use the term percent for specific numbers, as in "15 percent of the population." Use the term **percentage** as a general term, as in "a small percentage of the population."

One percent (1%) is the same as $\frac{1}{100}$ of a whole. A fraction with 100 in the denominator can be expressed as a percentage by simply taking the numerator and putting a percent sign after it. 15 out of 100 is 15 percent, written as 15%. Other fractions with a factor of 100 in the denominator can be renamed as equivalent fractions of 100 and then expressed as a percentage.

$$\frac{15}{100} = 15\%$$

$$\frac{7}{10} = \frac{70}{100} = 70\%$$

To express or rename a decimal number as a percentage, we can first rename it as a fraction of 100. Essentially, we are multiplying the decimal by 100 to express it as a percentage and thus we move the decimal over two places to the right. Since there are 100 percentage parts in a whole, converting a decimal to a percentage is similar to converting measurement, with % as the smaller unit, the whole as the larger unit, and 100 as the conversion unit.

$$0.25 = \frac{25}{100} = 25\%$$

$$0.25 \times 100\% = 25\%$$

To rename a percentage as a decimal number, we can first write it as a fraction of 100, and then rename that fraction as a decimal. Essentially, we are dividing the percentage by 100 (100 parts in a whole) to express it as a decimal and thus we move the decimal point two places to the left.

$$25\% = \frac{25}{100} = 0.25$$

To rename a percentage as a fraction in its simplest form, we can write it as a fraction of 100, and then simplify.

$$25\% = \frac{25}{100} = \frac{1}{4}$$

(1) Interpret percentages

Discussion

Concept p. 61

Have your student note that there are 100 seats, and 55 of them are occupied. Point out that we can write the number of occupied seats as a fraction of the total number of seats, $\frac{55}{100}$. We can also write it as a decimal, 0.55 seats.

Tell her that if the total is divided in 100 parts, we can write a fraction of 100 in a new way, as a *percentage*. Write 55% and read it; fifty-five *percent* of the total number of seats are occupied. Tell her that the % symbol is read as "percent" and means "out of 100."

Ask your student how many and then what percentage of the seats are unoccupied. 45% of the seats are unoccupied.

Discuss other ways in which your student might have heard the term percent. For example: "There is a 90% chance of rain today." "The top 1% of the population controls 43% of the wealth in the United States." "About 78% of a newborn baby is water." "Sale! 30% off all sweaters."

> $\frac{55}{100}$ of the seats are occupied.
>
> 0.55 of the seats are occupied.
>
> 55% of the seats are occupied.
>
> $\frac{55}{100} = 0.55 = 55\%$
>
> $\frac{45}{100} = 45\%$
>
> 45% of the seats are not occupied.

Tasks 1-2, p. 62

These tasks use 10 by 10 grids to illustrate the connection between percentage and fraction.

> 1. 27%
>
> 2. (a) 67%   (b) 50%   (c) 9%   (d) 100%

They provide a good visual image of the meaning of percentage. If needed, you can expand on this lesson by having your student color in the 10 by 10 grids on appendix p. a8 for percentages you name.

Task 2: Ask your student to write each amount as a fraction as well as a percentage. Point out that the fraction that is shaded in 2(b) can also be written as tenths rather than hundredths. $\frac{5}{10}$ of the whole is shaded. So if we are given a fraction with 10 in the denominator, we can think of the equivalent fraction of 100 to find the part out of 100 and write that as a percentage. 2(d) illustrates that 1 whole is 100%.

Point out that as with fractions, percentages are always of some whole. If the whole is not named, it is simply 1 whole, such as one square as in these tasks. (Later, your student will find percentages of a set of other than 100 items.)

Practice

Tasks 3-4, p. 62

Workbook

Exercise 1, pp. 47-48 (answers p. 74)

> 3. (a) 33%   (b) 20%   (c) 5%
>
> 4. (a) 23%   (b) 45%   (c) 36%   (d) 75%
>   (e) 40%   (f) 70%   (g) 30%   (h) 50%
>
> $\frac{3}{10} = \frac{30}{100}$

(2) Express percentages as fractions or decimals

Activity

Write 0.61 and ask your student to write it as a fraction and then as a percentage. Point out that decimal numbers are also fractions, so we can say 61 hundredths of the whole, or 0.61 of the whole, or 61% of the whole.

$$0.61 = \frac{61}{100} = 61\%$$

You can point out that converting a decimal to a percentage is very similar to converting meters to centimeters, since there are 100 centimeters in 1 meter, and 100 "percents" in 1 "whole." So we can convert the decimal to a percentage by simply multiplying it by the "conversion factor" 100, or moving the decimal point two places to the right.

$$0.61 \text{ whole} \times 100\% = 61\%$$

Ask your student to convert 0.7 to a percentage.

$$0.7 = \frac{70}{100} = 70\%$$

Write 37% and ask your student to convert it to a decimal. We can do this by first expressing it as the fraction 37 out of 100 and converting that to the decimal.

$$37\% = \frac{37}{100} = 0.37$$

You can tell your student that if we think of percentage as a unit of a whole, there are 100 units in the whole. To convert from the smaller unit, percent, to the larger one, whole, we divide by the conversion factor, 100, or move the decimal point two places to the left.

$$37\% = 37 \times \frac{1}{100} = 0.37 \text{ whole}$$

Ask your student to express 30% as a fraction and as a decimal. Then ask him to put the fraction in simplest form.

$$30\% = \frac{30}{100} = 0.3$$

$$30\% = \frac{30}{100} = \frac{3}{10}$$

Ask your student to express 75% as a fraction in simplest form.

$$75\% = \frac{75}{100} = \frac{3}{4}$$

Practice

Tasks 5-10, p. 63

Workbook

Exercise 2, pp. 49-51 (answers p. 74)

Reinforcement

Extra Practice, Unit 9, Exercise 1, pp. 185-190

Test

Tests, Unit 9, 1A and 1B, pp. 75-79

5. 35%

6. (a) 7%   (b) 2%   (c) 85%   (d) 70%

7. 0.43

8. (a) 0.28   (b) 0.88   (c) 0.3   (d) 0.05

9. $\frac{2}{5}$

10. (a) $\frac{1}{10}$ (b) $\frac{4}{5}$ (c) $\frac{1}{4}$ (d) $\frac{3}{4}$

     (e) $\frac{1}{20}$ (f) $\frac{2}{25}$ (g) $\frac{1}{25}$ (h) $\frac{1}{50}$

Chapter 2 – Writing Fractions as Percentages

Objectives

♦ Express a fraction as a percentage.
♦ Solve word problems that involve finding the percentage of the whole.

Material

♦ 10 by 10 grids (appendix p. a8)
♦ Mental Math 10-11 (appendix)

Notes

In this chapter, students will learn to convert fractions which have a denominator other than 10 or 100 into a percentage. There are 3 methods.

Method 1

Find an equivalent fraction of 100 and then write the fraction as a percentage.

$$\frac{1}{4} = \frac{25}{100} = 25\%$$

This method can be used when the denominator is a factor of 100 so it is easy to find an equivalent fraction with a denominator of 100.

Method 2

Multiply the fraction by 100%.

We can think of a whole as divided into 100 parts or units, each with a value of 1%. $\frac{1}{4}$ of this whole is $\frac{1}{4}$ of 100 parts which is 25 parts of 100, or 25%.

$$\frac{1}{4} \times 100\% = 25\%$$

$$\frac{3}{\underset{2}{\cancel{8}}} \times \cancel{100}^{25}\% = 37.5\%$$

This method is easier to use when it is not convenient to find the equivalent fraction with a denominator of 100. Your student should simplify as much as possible before multiplying or dividing.

Method 3

Divide the fraction to change it to a decimal. Then write the decimal as a percentage.

$$\frac{1}{4} = 0.25 = 25\%$$

0.25 of the whole is 0.25 x 100% of the whole. Thus we can convert a decimal to a percentage by multiplying it by 100. This is the same as moving the number's decimal point over two places to the right

This last method is not emphasized in this curriculum. It is similar to the second method except that there are now two steps; first dividing by a fraction, and then multiplying by 100 to get the percentage. The second method, where both steps are combined into one equation, offers more opportunities for simplifying the calculations and performing mental computation.

Your student should be able to easily find the percent of each of the following fractions listed at the right. Once she knows the percent equivalents for $\frac{1}{2}$, $\frac{1}{4}$, $\frac{1}{5}$, and $\frac{1}{10}$ she can easily calculate the others.

Your student will be solving word problems involving percentages. In some of these word problems, the total is divided up into two or three parts and he is asked to find the percentage of the unknown parts. For example:

⇒ In a collection of 50 red, blue, and green marbles, 10 are red and 25 are blue. What percentage of the marbles are green?

This can be solved by first finding the percentage of red marbles and the percentage of blue marbles, and then subtracting both from 100%, or by finding the total number of red and green marbles and then the percentage of both combined and subtracting from 100, or by finding the number of green marbles first and then the percentage of green marbles.

When doing word problems that involve finding the percentage part of the whole, make sure your student knows which amount is the whole. You may want to call it the *base*. It will be the amount that goes in the denominator of the fraction. At this level, it is always the larger amount, but in *Primary Mathematics* 6A students will be using percentages that are greater than 100%, so the whole may be the smaller number. Therefore, your student will not be able to simply look for the largest number as the whole.

$$\frac{1}{2} = \frac{2}{4} = \frac{5}{10} = 50\% \qquad \frac{4}{5} = \frac{8}{10} = 80\%$$

$$\frac{1}{4} = 25\% \qquad \frac{1}{10} = 10\%$$

$$\frac{3}{4} = 3 \times 25\% = 75\% \qquad \frac{3}{10} = 30\%$$

$$\frac{1}{5} = \frac{2}{10} = 20\% \qquad \frac{7}{10} = 70\%$$

$$\frac{2}{5} = \frac{4}{10} = 40\% \qquad \frac{9}{10} = 90\%$$

$$\frac{3}{5} = \frac{6}{10} = 60\%$$

Red marbles: $\frac{10}{50} \times 100\% = 20\%$

Blue marbles: $\frac{25}{50} \times 100\% = 50\%$

Green marbles: $100\% - 20\% - 50\% = 30\%$

Or

Red + blue marbles: $10 + 25 = 35$

$$\frac{35}{50} \times 100\% = 70\%$$

Green marbles: $100\% - 70\% = 30\%$

Or

Green marbles: $50 - 10 - 25 = 15$

$$\frac{15}{50} \times 100\% = 30\%$$

30% of the marbles are green.

(1) Express fractions (denominator < 100) as percentages

Discussion

Concept p. 64

Method 1: Since percentage is parts per hundred, in order to rename $\frac{3}{4}$ as a percentage we can find an equivalent fraction with 100 in the denominator. The numerator is then the same as the percentage. Tell your student to imagine the whole fence as divided into 100 parts. Each fourth is 25 parts of the hundred, and so $\frac{3}{4}$ is 3 x 25, 75 parts of the hundred, or $\frac{75}{100}$. He painted 75% of the fence.

$$\frac{3}{4} = \frac{75}{100} = 75\%$$

Method 2: Tell your student to again think of the wall as divided into 100 parts. Since percentage means the number of parts out of 100, each part is 1%, and the total is 100%. To find out how many of these parts there are in $\frac{3}{4}$ of the wall, we find $\frac{3}{4}$ of the total 100 parts, or $\frac{3}{4}$ x 100%. This is similar to conversion of measurement; to go from $\frac{3}{4}$ of 1 meter to $\frac{3}{4}$ of 100 cm we multiply by 100. $\frac{3}{4}$ of 1 meter is the same as 75 cm; similarly $\frac{3}{4}$ of 1 whole of 100% is the same as 75%. Remind your student to always try simplifying as much as possible first.

$$\frac{3}{4} \times 100\% = 75\%$$

$$\frac{3}{\cancel{4}_1} \times \cancel{100}^{25}\% = (3 \times 25)\%$$
$$= 75\%$$

Optional: Discuss a third method. Since we can convert fractions to decimals and decimals to percentages, we can first convert $\frac{3}{4}$ to a decimal.

$$\frac{3}{4} = 3 \div 4 = 0.75 = 75\%$$

Tasks 1-3, p. 65

Task 1: This task illustrates the first method for converting a fraction to a percentage.

Task 2: This task shows both methods. You can illustrate this task by drawing a 5 x 5 square grid and shading in 7 squares. Then divide each square into 4 parts. There are now 100 equal parts. For method 1 we find the equivalent fraction. For method 2 we find the number of 1% parts that are shaded by finding the fraction of the set of 100.

Task 3: Have your student use both methods to find the percentage.

1. (a) 40%
   (b) 50%

2. 28%   28%

3. 70%

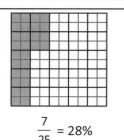

$$\frac{7}{25} = 28\%$$

Practice

Task 4, p. 65

4. (a) 25%	(b) 40%	(c) 80%	(d) 45%
(e) 65%	(f) 24%	(g) 56%	(h) 82%

Your student may be able to do most of these mentally. Since the denominators are all factors of 100, all he needs to do is determine what number to multiply the denominator by to get 100, and then multiply the numerator by that number.

Activity

Write the fraction $\frac{24}{40}$ and ask your student to try both methods to convert the fraction to a percentage. 40 is not a factor of 100. To use the first method, we could simplify and then find the equivalent fraction. For the second method, there are several ways to simplify the fraction of 100 before multiplying.

$$\frac{24}{40} = \frac{12}{20} = \frac{6}{10} = \frac{60}{100} = 60\%$$

$$\frac{24}{40} \times 100\% = \frac{24^{6}}{4_{1}} \times 10\%$$
$$= 60\%$$

Write the fraction $\frac{3}{8}$ and ask your student to express it as a percentage. The first method will not work as well. 8 is not a factor of 100 and the fraction cannot be simplified. The second method works better when the denominator is not a factor of 100. We can also convert the fraction to a decimal (using division) and then express the decimal as a percentage. Have her try this method and ask her which method she finds easier. Using fractions of 100 and simplifying gives an easier division problem that can be done mentally.

$$\frac{3}{8_{2}} \times 100^{25}\% = \frac{75}{2}\%$$
$$= 37.5\%$$
$$\text{or } 37\frac{1}{2}\%$$

$$3 \div 8 = 0.375 = 37.5\%$$

Repeat with $\frac{14}{60}$. There are several methods your student could use; multiply the fraction by 100% and simplify, divide 14 by 60 to get the decimal first, simplify it to $\frac{7}{30}$ and then divide. The decimal equivalent is a non-terminating decimal. Tell him that percentages can be rounded to a specified number of decimal places as well, but when the decimal is non-terminating he should generally express the percentage as a mixed number.

$$\frac{14}{60} \times 100\% = \frac{14^{7}}{6_{3}} \times 10\%$$
$$= \frac{70}{3}\%$$
$$= 23\frac{1}{3}\%$$

Write $\frac{49}{50}$ and ask your student to rename it as a percentage mentally. One method is to double 49. Another is to note that 49 is one less than 50, and subtract 1% of 50 from 100%.

$$\frac{49}{50} = \frac{49 \times 2}{50 \times 2} = 98\%$$

$$\frac{49}{50} = 1 - \frac{1}{50}$$
$$= 100\% - 2\%$$
$$= 98\%$$

Workbook

Exercise 3, pp. 52-53 (answers p. 74)

Reinforcement

Mental Math 10

(2) Express fractions (denominator > 100) as percentages

Discussion

Task 5, p. 66

5. 60%   60%

In this task, the whole is divided into 300 parts. To find the percentage of the colored part, we first write it as a fraction. Discuss the two methods shown for finding the percentage.

Method 1: Find an equivalent fraction with a denominator of 100. We need to divide both the numerator and denominator by 3.

Method 2: Find the fraction of 100%. We can simplify the equation in several ways.

You can have your student think of this as "merging" 3 units together in order to have 100 units, each of which is 1%.

Task 6, p. 66

6. 49%
   49%

$$\frac{98}{200} = \frac{49}{100} = 49\%$$

$$\frac{98}{200}\!\!\!\!\diagup_2 \times \overset{1}{\cancel{100}}\% = 49\%$$

First express 98 out of 200 as a fraction. You can have your student try both methods for converting the fraction to a percentage.

Practice

Tasks 7, p. 66

7. (a) 4%   (b) 18%   (c) 20%   (d) 43%
   (e) 10%   (f) 32%   (g) 4%   (h) 51%

Your student can probably do these mentally, dividing the numerator by the hundreds digit.

Activity

Point out that the first method can be used if the denominator is an easy multiple of 100.

Write $\frac{60}{240}$ and ask your student to write it as a percentage. Point out that although the denominator is not a factor or multiple of 100, this fraction can be simplified to an easy fraction. A good rule is to always try simplifying the fraction first.

$$\frac{60}{240} = \frac{1}{4} = 25\%$$

Write $\frac{67}{320}$ and ask your student to convert it to a percentage. This fraction cannot be simplified. Dividing it to find the decimal is a chore. If we multiply by 100%, we can do some simplification, and even convert to a mixed number before multiplying to make calculations easier.

$$\frac{67}{32\cancel{0}} \times 10\cancel{0}\%$$

$$= \frac{67}{32} \times 10\%$$

$$= \left(\frac{64}{32} + \frac{3}{32}\right) \times 10\%$$

$$= \left(2 + \frac{3}{32}\right) \times 10\%$$

$$= \left(20 + \frac{30}{32}\right)\%$$

$$= 20\frac{15}{16}\%$$

Workbook

Exercise 4, pp. 54-55 (Answers p. 74)

Reinforcement

Mental Math 11

(3) Solve word problems

Discussion

Tasks 8-11, p. 67

Task 8: Have your student find the percentage that is unshaded as well. Point out that she can find the percentage that is unshaded by subtracting the percentage that is shaded from 100%, or she can simply use the percentage scale.

Task 10: Ask your student to identify the whole (25).

Task 11: Ask your student to identify the whole ($750). Point out that since we do not have to first find the percentage of the money he spent, we can find the percentage he saved by subtracting the money he spent from his total money first.

Practice

Ask your student to find the following. If the percentage is not a whole number, he should generally express it as a mixed number, especially if it is a non-terminating decimal.

1.   20 minutes as a percentage of an hour.

2.   5 months as a percentage of a year.

3.   5 months as a percentage of 5 years.

4.   35 g as a percentage of 2 kg.

Workbook

Exercise 5, pp. 56-57 (answers p. 74)

---

8. (a) 10%
   (b) 50%
   (c) 80%
   (d) 70%
   (e) 40%

9. (a) 75%
   (b) 25%
       25%

10. (a) $\frac{7}{25}$ x 100% = (7 x 4)% = **28%**

    28% of the children are boys.

    (b) 100% − 28% = **72%**

    72% of the children are girls.

11. He saved $750 − $300 = $450.
    Percentage saved:

    $\frac{450}{750}$ x 100% = $\frac{3}{5}$ x 100% = (3 x 20)% = **60%**

    Or: Percentage spent:

    $\frac{300}{750}$ x 100% = $\frac{10}{25}$ x 100% = (10 x 4)% = 40%

    Percentage saved: 100% − 40% = 60%
    He saved 60% of his money.

---

1. $\frac{20}{60}$ x 100% = $33\frac{1}{3}$%

2. $\frac{5}{12}$ x 100% = $41\frac{2}{3}$%

3. $\frac{5}{5\times12}$ = $\frac{1}{12}$ x 100% = $8\frac{1}{3}$%

4. $\frac{35}{2000}$ x 100% = $1\frac{3}{4}$% or 1.75%

(4) Practice

Practice

Practice A, p. 68

Reinforcement

Extra Practice, Unit 9, Exercise 2, pp. 191-196

Test

Tests, Unit 9, 2A and 2B, pp. 81-84

---

1. (a) 5%   (b) 36%   (c) 60%   (d) 5%

2. (a) 63%   (b) 5%   (c) 20%   (d) 50%

3. (a) $\frac{23}{50}$ (b) $\frac{1}{20}$ (c) $\frac{7}{100}$ (d) $\frac{4}{5}$

4. (a) 0.15   (b) 0.41   (c) 0.09   (d) 0.5

5. $\frac{15}{100}$ = **15%**

   15% of the oranges are rotten.

6. Number of red marbles: $100 - 37 = 63$

   $\frac{63}{100}$ = **63%**

   63% of the marbles are red.

7. $\frac{60}{100}$ = $\frac{3}{5}$

   The team won $\frac{3}{5}$ of the games.

8. $100\% - 70\% =$ **30%**

   30% of the tank is not filled.

9. $\frac{4}{5} \times 100\% =$ **80%**

   80% of the books are fiction books.

10. $\frac{14}{50} \times 100\% =$ **28%**

   28% of the vehicles are motorcycles.

11. Number of adults: $1500 - 450 = 1050$

   $\frac{1050}{1500} \times 100\% =$ **70%**

   70% of the participants were adults.

12. Flour for pineapple tarts: $5\text{ kg} - 2\text{ kg} = 3\text{ kg}$

   $\frac{3}{5} \times 100\% =$ **60%**

   60% of the flour was used to make the tarts.

Workbook

Exercise 1, pp. 47-48

1. (a) 7%   (b) 15%   (c) 29%
   (d) 26%   (e) 38%   (f) 28%

2. Check answers.

3. 87%   5%
   16%   71%
   68%   50%
   99%   100%

4. 7   1
   43   99
   100   100
   100   100

Exercise 2, pp. 49-51

1. (a) 15%   (b) 86%
   (c) 40%   (d) 90%
   (e) 47%   (f) 12%
   (g) 4%   (h) 50%
   (i) 75%   (j) 6%

2. (a) 0.24   (b) 0.37
   (c) 0.78   (d) 0.06
   (e) 0.62   (f) 0.53
   (g) 0.1   (h) 0.07
   (i) 0.8   (j) 0.9

3. (a) $\frac{11}{50}$ (b) $\frac{9}{20}$

   (c) $\frac{24}{25}$ (d) $\frac{13}{25}$

   (e) $\frac{3}{50}$ (f) $\frac{2}{5}$

   (g) $\frac{9}{10}$ (h) $\frac{2}{25}$

   (i) $\frac{3}{4}$ (j) $\frac{1}{2}$

Exercise 3, pp. 52-53

1. (a) 50%   (b) 18%
   (c) 85%   (d) 48%
   (e) 60%   (f) 60%
   (g) 16%   (h) 25%
   (i) 24%   (j) 30%

2. (a) $\frac{8}{40} = \frac{2}{10} = $ **20%**

   (b) $\frac{40}{80} = \frac{1}{2} = $ **50%**

(c) $\frac{15}{50} = \frac{3}{10} = $ **30%**

(d) $\frac{7}{20} = \frac{35}{100} = $ **35%**

(e) $\frac{24}{40} = \frac{6}{10} = $ **60%**

Exercise 4, pp. 54-55

1. (a) $\frac{186}{200} = \frac{93}{100} = $ **93%**

   (b) $\frac{39}{300} = \frac{13}{100} = $ **13%**

   (c) $\frac{96}{400} = \frac{24}{100} = $ **24%**

   (d) $\frac{235}{500} = \frac{47}{100} = $ **47%**

   (e) $\frac{122}{200} = \frac{61}{100} = $ **61%**

2. $\frac{9}{20}$ x 100% = **45%**
   45% of the stamps were mailed to Canada.

3. $\frac{24}{80}$ x 100% = **30%**
   30% of them are 6th grade students.

4. $\frac{64}{200} = \frac{32}{100} = $ **32%**
   32% of the units are 3-bedrooms.

Exercise 5, pp. 56-57

1. (a) $\frac{24}{50}$ x 100% = **48%**
   48% of the them were chocolate.

   (b) 100% − 48% = **52%**
   52% of them are sugar cookies.

2. (a) $\frac{32}{80}$ x 100% = **40%**
   He spent 40% of his money on books.

   (b) 100% − 40% = **60%**
   He had 60% of his money left.

3. (a) $\frac{120}{400}$ x 100% = **30%**
   30% of the seats are occupied.

   (b) 100% − 30% = **70%**
   70% of them are not occupied.

4. (a) $\frac{85}{125}$ x 100% = **68%**
   68% of the swimmers are female.

   (b) 100% − 68% = **32%**
   32% of them are male.

Chapter 3 – Percentage of a Quantity

Objectives

♦ Find the value for a percentage of a whole.
♦ Solve word problems that involve percentage of a whole.
♦ Solve word problems that involve percentage tax, interest, discount, increase, decrease.

Material

♦ Mental Math 12 (appendix)

Vocabulary

♦ Income tax
♦ Sales tax
♦ Interest
♦ Discount
♦ Percent increase
♦ Percent decrease

Notes

In this chapter your student will learn to find the value for a percentage of the whole. There are three methods: the fraction method, the unitary method, and the decimal method.

Fraction method

Convert the percentage to a fraction and then find the fraction of the whole.

In the example at the right, we can simplify by crossing out equal numbers of 0's in the numerator or whole number and in the denominator even if one 0 comes from the numerator of the fraction, and the other from the whole number.

Unitary method

Find the value of 1% by dividing the whole by 100, and then multiply to find the value of more than 1%.

When this method is done in two separate steps, it sometimes involves multiplying a decimal number by a whole number. To simplify calculations and allow for the most opportunity to simplify the fractions, leave the final computation until the end.

When the percentage we want to find is a multiple of 10, we can first find the value for 10% instead of 1%. However, always finding 1% first has a conceptual advantage when starting out; it keeps your student aware of the underlying information and less likely to get confused by the process into just pushing numbers around.

Decimal method

Convert the percentage to a decimal and multiply the decimal by the whole.

Find 40% of 180.

Fraction method:

$$40\% \text{ of } 180 = \frac{40}{100} \times 180 = 72$$

$$\frac{40 \times 180}{100} = \frac{4 \times 10 \times 18 \times 10}{10 \times 10}$$

$$\frac{4\cancel{0}}{1\cancel{0}\cancel{0}} \times 18\cancel{0} = 4 \times 18 = 72$$

Unitary method:

100% of 180 = 180

$$1\% \text{ of } 180 = \frac{1}{100} \times 180 = \frac{180}{100} = 1.8$$

40% of 180 = 1.8 x 40 = 72

Or

$$1\% \text{ of } 180 = \frac{180}{100}$$

$$40\% \text{ of } 180 = \frac{18\cancel{0}}{1\cancel{0}\cancel{0}} \times 4\cancel{0} = 72$$

Or

10% of 180 = 18
40% of 180 = 18 x 4 = 72

Decimal method:

40% of 180 = 0.40 x 180 = 72

The decimal method has the least chance for mental calculation since there is no opportunity to simplify fractions first. In this curriculum, the decimal method isn't used at this level, except as might arise from finding 1% of a number. Instead, the fraction method will be used primarily. The unitary method will be used more in *Primary Mathematics* 6.

A fraction can often be simplified in different ways, as shown in the example at the right. Simplifying a fraction as much as possible before performing other operations will make the multiplication easier.

In discussion emphasize which number is the whole. At this level, it will generally be the largest number. In *Primary Mathematics* 6 students will be finding percentages greater than 100%, in which case the whole is not the larger number.

Find 15% of 40.

$$\frac{15}{100} \times 40 = \frac{15 \times 4^2}{10_5} = \frac{15^3 \times 2}{5_1} = 6$$

$$\frac{15^3}{100_{20}} \times 40 = \frac{3 \times 40^2}{20_1} = 6$$

Your student will learn to solve problems involving tax, interest, discount, increase, and decrease. He can become familiar with these terms through sufficient discussion.

Income tax is the percentage of your income the government receives.

Sales tax is the percentage of the purchase price of goods the government receives and is added to the store price on purchase.

Interest is the amount of money a bank pays you to have your money in its bank, or you pay the bank to borrow their money.

Discount is the percentage of the total price that is subtracted from the total price when an item for sale is discounted.

Increases and **decreases** of population or cost or other numerical values are often given as a percentage increase or decrease compared to the original amount.

(1) Find the vale for a percentage of a quantity

Discussion

Concept p. 69

You may want to first write the problem on a whiteboard or paper so you can discuss it with your student to give her an opportunity to apply concepts she has already learned to this new situation.

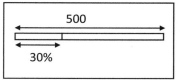

Tell your student that in this problem we need to find part of the whole and ask him what the whole is (500). If not using the text yet, diagram the information with a bar for the whole. Then ask him to mark the part we need to find. 30% is a little less than a third. Let him suggest ways to find 30% of the whole. Then discuss the two methods.

Method 1: Think of the total as divided into 100 equal units. Each unit is 1%. If we find the value of 1 unit, we can then find the value of any number of units, including 30 units, which would be 30%.

$$100 \text{ units} = 500$$

$$1 \text{ unit} = \frac{500}{100} = 5$$

$$30 \text{ units} = 5 \times 30 = 150$$

Method 2: Find a fraction of a whole. We rename the decimal as a fraction and find the fraction of the whole.

$$30\% = \frac{30}{100}$$

$$\frac{30}{100} \times 500 = 150$$

$$\frac{30}{1\cancel{0}\cancel{0}} \times 5\cancel{0}\cancel{0} = 30 \times 5 = 150$$

Discuss different ways to simplify the problem before multiplying. Usually the easiest thing to do is to simplify tens or hundreds first. Since we are dividing 30 x 500 by 100, we can show the simplification (dividing both numerator and denominator by 100) by crossing out the same number of 0's in the numerator and the denominator. Make sure your student understands we are simplifying the fraction, not throwing away 0's. Then we can simplify further if needed.

If you did not use the picture in the text yet, have your student look at the solutions in the textbook now.

Tell your student that it is always a good idea to check if an answer is reasonable. 30% is about a third of 500. Is 150 a reasonable answer? Since a third of 600 is 200, it is reasonable. If we forget something in the calculation, like crossing off the wrong number of 0's, or forgetting to multiply by tens after multiplying by 3, we could get an answer of 15, which is not a reasonable answer.

Expand on the problem with the following questions.

⇒ 25% of the people at the concert were men. How many men were there?

Point out that she should know that 25% is a fourth, and can simply find a fourth of 500.

⇒ 75% of the people wore T-shirts with the band's name on it. How many wore these T-shirts?

If she remembers that 75% is three fourths, she can simply divide the total into 4 units and find 3 of them.

Similarly, in the original problem, if your student has not already suggested it, point out that since 30% is 3 tenths, we could divide the bar into ten units and find the value of 3 of them.

⇒ 15% of the people brought food to eat. How many brought food?

Since we already know what 10% is, 15% is just 10% plus half of 10%.

⇒ 18% of the people left the concert early. How many stayed until the end?

In this case, it is probably easier to use one of the two standard methods than to come up with a shortcut. Tell her she can use shortcuts whenever she wants, but if she does not quickly think of a good shortcut to use then just use of the two methods.

Tasks 1-2, pp. 69-70

You can discuss different strategies for solving these.

Practice

Task 3, p. 70

Your student may be able to solve some of these mentally.

Workbook

Exercise 6, pp. 58-59 (answers p. 81)

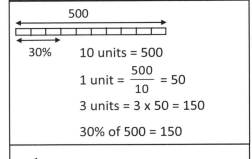

500

25%

4 units = 500

1 unit = $\dfrac{500}{4}$ = 125

25% of 500 is 125.

75% of 500 = 125 x 3 = 375

500

30%

10 units = 500

1 unit = $\dfrac{500}{10}$ = 50

3 units = 3 x 50 = 150

30% of 500 = 150

$\dfrac{1}{2}$ of 10% of 500 = 25

15% of 500 = 50 + 25 = 75

18% of 500 = ?

$\dfrac{18}{1\cancel{0}\cancel{0}}$ x 5$\cancel{0}\cancel{0}$ = 18 x 5 = 90

---

1. 1% of 120 = 1.2
   90% of 120 = 1.2 x 90 = **108**

   or: 1% of 120 = $\dfrac{120}{100}$

   90% of 120 = $\dfrac{120}{100}$ x 90 = 108

   or: 90% of 120 = $\dfrac{90}{100}$ x 120 = 108

   or: 10% of 120 = 12
   90% of 120 = 12 x 9 = 108
   108 students passed the test.

2. 1% of $800 = $8
   3% of $800 = $8 x 3 = **$24**

   or: 3% of $800 = $\dfrac{3}{100}$ x $800 = $24

   The sales tax was $24.

3. (a) 15   (b) 16   (c) 10 kg
   (d) 10 m   (e) 31.5 km   (f) 300 g

(2) Solve word problems involving percentage of a quantity

Discussion

Tasks 4-5, p. 70-71

Discuss these two problems with your student. You may want to write them on paper or whiteboard and have him explain his solution before looking at the method given in the textbook.

Task 5: Two methods are given. In the first method we first find the percent of the members that are adults, and then use that percentage to find the number of adults. In the second method we first use the percentage of children to find the number of children, and then use that to find the number of adults.

> 4. (a) 40%
>   (b) $200
>
> 5. 352
>   352
>   352
>   352

Practice

Practice B, problems 1-5, p. 73

Workbook

Exercise 7, pp. 60-61 (answers p. 81)

Reinforcement

Mental Math 12

> 1. (a) 6.56   (b) 123.2   (c) 322.4 m
>
> 2. 7% of 60 m^2 = $\dfrac{7}{100}$ x 60 m^2 = **4.2 m^2**
>   The area of the pond is 4.2 m^2.
>
> 3. 10% of 50 = 5
>   90% of 50 = 9 x 5 = **45**
>   She spelled 45 of them correctly.
>
> 4. 100% – 55% = 45%
>   45% of 20 = $\dfrac{45}{100}$ x 20 = **9**
>   There are 9 male workers.
>
> 5. 10% of $1350 = $135
>   30% of $1350 = $135 x 3 = **$405**
>   She saves $405 each month.

(3) Solve word problems involving interest and discounts

Discussion

Tasks 6-9, pp. 71-72

Task 6: Explain that interest is the amount of money a bank gives you for letting it keep your money. It is usually figured as a percentage of the amount you have in the bank. If you borrow money from the bank, you pay the bank interest, which is a percentage of what you owe the bank. (You can go into additional explanations if you want about the advantages and disadvantages of borrowing money.)

Task 7: Explain that a discount is how much is taken off of the original price and is given as a percentage of the original price. You may want to discuss examples in local stores; discounts may be given in sale signs as percent off. Point out that the original price is what determines how much off you actually get. For example, 10% off of $40 is more than 10% off of $20. (You may want to discuss ways stores get you to come buy; such as raising the original price before announcing a sale.)

Tasks 8-9: Increases and decreases are often given as percentage of the original amount. Percent increase or decrease is often used for increases or decreases in money or population.

6. $81 $2781 $2781
7. $108 $792 $792
8. $120 $1620 $1620
9. 20 380 380

Practice

Practice B, problems 6-12, p. 73

Workbook

Exercises 8-9, pp. 62-64 (answers p. 81)

Reinforcement

Extra Practice, Unit 9, Exercise 3, pp. 197-200

Test

Tests, Unit 9, 3A and 3B, pp. 85-88

6. 5% of 720 = 36 (5% is half of 10%, so the increase is half of 72.) 720 + 36 = **756** There are 756 members this year.
7. Tax = 3% of $50 = $1.50 $50 + $1.50 = **$51.50** She paid $51.50 for the swimsuit.
8. 10% of $190 = $19 30% of $190 = $19 x 3 = $57 $190 – $57 = **$133** The sale price of the camera was $133.
9. 3% of $3500 = $105 $3,500 + $105 = **$3605** She had $3605 in the bank after 1 year.
10. 100% – 40% = 60% 60% of 15 = **9** 9 arrows did not hit.
11. 100% – 30% – 40% = 30% 30% of 280 = **84** There are 84 adult members.
12. 100% – 10% – 75% = 15% 15% of 200 = **30** There are 30 spaces for motorcycles.

Workbook

Exercise 6, pp. 58-59

1. (a) $\dfrac{4}{100} \times 300 = $ **12** (b) $\dfrac{72}{100} \times 15 = $ **10.8**

   (c) $\dfrac{30}{100} \times \$94 = $ **\$28.20** (d) $\dfrac{5}{100} \times \$250 = $ **\$12.50**

   (e) $\dfrac{25}{100} \times 240$ m $= $ **60 m** (f) $\dfrac{80}{100} \times 25$ kg $= $ **20 kg**

2. 55% of $85 = \dfrac{55}{100} \times \$85 = $ **\$46.75**

   She paid \$46.75 for electricity.

3. 25% of 48 $= \dfrac{25}{100} \times 48 = $ **12**

   12 accidents happened on the freeways.

4. 30% of $750 = \dfrac{30}{100} \times \$750 = $ **\$225**

   She gives \$225 to charity.

Exercise 7, pp. 60-61

1. Green apples: 100% − 40% = 60%

   60% of 55 $= \dfrac{60}{100} \times 55 = $ **33**

   There are 33 green apples in the box.

2. Kept: 100% − 30% = 70%

   70% of $840 = \dfrac{70}{100} \times \$840 = $ **\$588**

   He kept \$588.

3. Spent: 100% − 15% = 85%

   85% of $1200 = \dfrac{70}{100} \times \$12 = $ **\$1020**

   He spends \$1020.

4. Incorrect answers: 100% − 82% = 18%

   18% of 750 $= \dfrac{18}{100} \times 750 = $ **135**

   He answered 135 questions incorrectly.

Exercise 8, pp. 62-63

1. (a) 6% of $1800 = \dfrac{6}{100} \times \$1800 = $ **\$108**

   She earns \$108 in interest after a year.

   (b) \$1800 + \$108 = **\$1908**

   She will have \$1908 in the bank after a year.

2. Interest = 8% of $2800 = \dfrac{8}{100} \times \$2800 = \$224$

   \$2800 + \$224 = **\$3024**
   She has to pay \$3024.

3. (a) 20% of $60 = \dfrac{20}{100} \times \$60 = $ **\$12**

   The discount was \$12.

   (b) \$60 − \$12 = **\$48**

   The selling price was \$48.

4. 25% of $15 = \dfrac{25}{100} \times \$15 = \$3.75$

   \$15 − \$3.75 = **\$11.25**
   The selling price is \$11.25

Exercise 9, p. 64

1. 12% of $300 = \dfrac{12}{100} \times 300 = $ **\$36**

   He has to pay \$36 more each month.

2. Increase = 4% of 1500 $= \dfrac{4}{100} \times 1500 = 60$

   1500 + 60 = **1560**
   The number of workers after the increase is 1560.

Review 9

Review

Review 9, pp. 74-75

Workbook

Review 9, pp. 65-68 (answers p. 83)

Tests

Tests, Units 1-9, Cumulative Tests A and B, pp. 89-96

1. (a) 7.28 km   (b) 3.01 kg
   (c) 0.28 ℓ   (d) 30.46 cm

2. (a) 69.409   (b) 107.809

3. (a) 0.8   (b) 0.4   (c) 3.3   (d) 3.6

4. (a) 49.89   (b) 147.10   (c) 2209.52

5. 3^2 x 13

6. (a) 24   (b) 10

7. (a) 137.2   (b) 316.68   (c) 9.675

8. (a) 3^2 x 5^2 x 7   (b) 2^3 x 7 x 11^2 x 13

9. (a) $5\frac{7}{20}$ (b) $14\frac{14}{15}$ (c) $\frac{8}{9}$
   (d) $\frac{4}{15}$ (e) $\frac{7}{12}$ (f) 100

10. $\frac{20 \text{ min}}{180 \text{ min}} = \frac{1}{9}$

11. (a) 14%   (b) 6%

12. (a) 75%   (b) 75%   (c) 65%

13. (a) $1.80   (b) 45 g   (c) 42 m

14.

To show fifths and half, use 10 units. 1 unit is poured out.
1 unit = 500 ml
10 units = 5000 ml = **5 ℓ**
The capacity is 5 liters.

15. 3 units = 4 kg

    1 unit = $\frac{4}{3}$ kg

    2 units = $\frac{4}{3}$ kg x 2 = $\frac{8}{3}$ kg = **$2\frac{2}{3}$ kg**

    Half of the bag weighs $2\frac{2}{3}$ kg.

16. $\frac{186}{200}$ = **93%**

    93% of the families own computers.

17. 25% of $6 = $1.50
    $6 + $1.50 = **$7.50**
    The selling price was $7.50.

18. 15% of $180 = $27
    $180 − $27 = **$153**
    The selling price was $153.

19. $\frac{1}{4}$ x 40 cm = 10 cm

    Water in A: 24 cm x 10 cm x 10 cm = 2400 cm^3
    Height water in A adds to water in B:

    $\frac{2400 \text{ } cm^3}{30 \text{ cm} \times 20 \text{ cm}}$ = 4 cm

    New height of water in B: 16 cm + 4 cm = 20 cm

    $\frac{2}{3}$ of total height of B = 20 cm

    Total height of B = $\frac{20}{2}$ x 3 = **30 cm**

    Container B is 30 cm high.

Workbook

Review 9, pp. 65-68

1. (a) 1   (b) 1000

2. (a) 3.016
   (b) 3.601
   (c) 3.061

3. (a) $\dfrac{5}{3} + 4\dfrac{1}{8}$

   $= 1\dfrac{16}{24} + 4\dfrac{3}{24}$

   $= 5\dfrac{19}{24}$

   (b) $\dfrac{\cancel{2}^{1}}{7} \times \dfrac{5}{\cancel{8}_{4}}$

   $= \dfrac{5}{28}$

   (c) $\dfrac{5}{6} \div \dfrac{1}{3}$

   $= \dfrac{5}{\cancel{6}_{2}} \times \cancel{3}^{1}$

   $= 2\dfrac{1}{2}$

   (d) 13.3   (b) 12.0725

4. (a) 48%
   (b) 20%
   (c) 1%
   (d) 12%

5. (a) 57%
   (b) 25%
   (c) 15%
   (d) 3%

6. (a) $\dfrac{3}{4}$

   (b) $\dfrac{1}{4}$

   (c) $\dfrac{3}{50}$

   (d) $\dfrac{2}{5}$

7. (a) 6   (b) 25
   (c) 9   (d) 28

8. 72, 144

9. (a) 5 gal 3 qt
      = (5 x 4 qt) + 3 qt
      = **23 qt**
   (b) 4 ft 11 in.
      = (4 x 12 in.) + 11 in.
      = **59 in.**
   (c) 5 lb 13 oz
      = (5 x 16 oz) + 13 oz
      = **93 oz**
   (d) 3 qt 2 pt
      = (3 x 2 pt) + 2 pt
      = **8 pt**

10. $\dfrac{4}{5}$ qt $- \dfrac{1}{2}$ qt $= \dfrac{8}{10}$ qt $- \dfrac{5}{10}$ qt $= \mathbf{\dfrac{3}{10}}$ **qt**

11. $\dfrac{1}{4}$ x 5 ft $= \dfrac{5}{4}$ ft $= \mathbf{1\dfrac{1}{4}}$ **ft**

12. (a) $\dfrac{3}{4}$ lb = 12 oz; $\dfrac{13}{16}$ lb = 13 oz

    11 oz, $\dfrac{3}{4}$ lb, $\dfrac{13}{16}$ lb

    (b) 1.5 gal = 6 qt
    1.5 gal, 10 qt, 16 qt

    (c) $1\dfrac{2}{3}$ yd = 5 ft = 60 in.; $3\dfrac{1}{6}$ ft = (36 + 2) in. = 38 in.

    $\mathbf{3\dfrac{1}{6}}$ **ft, 39 in., $1\dfrac{2}{3}$ yd**

13. (a) $\dfrac{1}{2}$ x 8 in. x 15 in.
       = **60 in.²**
    (b) $\dfrac{1}{2}$ x 8 ft x 12 ft
       = **48 ft²**

14. $\dfrac{156}{200}$ x 100% = **78%**

15. 15% = $\dfrac{15}{100}$ = $\dfrac{3}{20}$

16. 1200 − 780 = 420

    $\dfrac{420}{1200}$ x 100% = **35%**

17. $\dfrac{12}{100}$ x $450 = $54

    $450 + $54 = **$504**

18. 90 − 36 = 54

    $\dfrac{54}{90}$ x 100% = **60%**

19. 500g − 200 g = 300 g

    $\dfrac{300}{500}$ x 100% = **60%**

20. Cost per bicycle: $7000 ÷ 50 = $140
    35% of $140 = **$49**

Unit 10 – Angles

Chapter 1 – Measuring Angles

Objectives

- Estimate and measure angles in degrees.
- Tell direction in relation to an 8-point compass.
- Determine the angle between various points on the compass.

Material

- Protractor
- Compass (directional)
- Folding meter stick or 2 strips of cardboard attached at one end with a brad, or similar construction toy
- Paper circle (appendix p. a9)

Vocabulary

- Clockwise
- Counterclockwise
- Angle
- Degree
- Protractor

- Right angle
- Acute angle
- Obtuse angle
- Reflex angle

Notes

Students learned how to measure angles, including angles greater than 180° in *Primary Mathematics* 4A. This is reviewed and reinforced in this chapter.

An **angle** is formed when two straight lines meet at a point. The point of intersection is the vertex of the angle, and the two lines are the sides of the angle. The size of an angle is determined by how much either line is turned about the point where they meet. It does not depend on the length of the two sides.

Angle b is larger than angle a.

The **degree** is derived from the Babylonian base-60 system. They may have assigned 360 degrees to a circle because they found that it took about 360 days for the sun to complete one year's circuit across the sky. 360 is conveniently divisible by 2, 3, 4, 5, 6, 8, 9, 10, 12, 15, 18, and 20, so the degree is a nice unit to use to divide the circle into an equal number of parts. 180 degrees is half of the way around a circle, 120 degrees is one third of a the way around a circle, 90 degrees is one fourth of the way around a circle, and so on. The abbreviation for degree is a superscript O (e.g., 90°).

In this chapter your student will be introduced to the angles of an 8-point compass. The angle between any two adjacent points on an 8-point compass is 45°. Your student will be finding the total degree of turning through various points of the compass by either finding multiples of 45°or adding 45° to 90°, 180°, or 270°. Along with counting right angles, this concretely introduces students to the additive property of angles and can help with estimating angles. Relating angles to compass points also emphasizes that the size of an angle is a measure of the degree of turning and reiterates the relationship between angles and circles using the perspective of being in the center of the circle looking towards the edge and turning.

(1) Review angles

Discussion

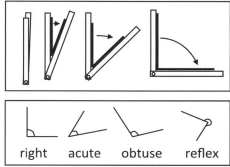

Use two strips of cardboard fastened with a brad at one end, or other suitable object, to discuss angles if needed. Remind your student that angles are formed when two lines cross each other or meet at a point. They are measured by the amount, or degree, of turning of one line relative to another line about the common point. The larger the angle, the more one side has to be turned away from the other. The size of the angle depends on the degree of turning. A complete turn is a circle. See if he remembers what one fourth of a turn around the circle is called (a *right angle*). Remind him that an angle smaller than a right angle is called an *acute angle*, an angle larger than one right angle but smaller than two right angles is called an *obtuse angle*, and an angle larger than two right angles is called a *reflex angle*. Point out that every angle actually has an opposite angle, and a reflex angle is opposite a smaller angle. Draw and mark some angles and ask him to identify them as right, acute, obtuse, or reflex.

right   acute   obtuse   reflex

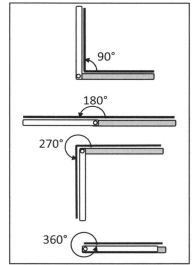

Provide your student with a protractor. Remind her that the customary unit for measuring angles is degrees and is abbreviated with a little superscript O. Ask her for the number of degrees in one full turn around the circle (360°). Form a right angle with the strips, which is a quarter turn, or draw a right angle. Ask her for the number of degrees in a right angle (90°). Point out that 90° is one fourth of 360°. Similarly, ask her for the number of right angles and the number of degrees in a half-turn (180°) and a three quarter turn (270°).

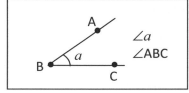

Remind your student that *angle* is abbreviated as ∠. An angle is named with a lowercase letter, or with three points named with uppercase letters, one on each arm and one at the vertex, with the one at the vertex in the middle.

Draw a 90° angle and then bisect it with a line. Ask your student for the size of an angle that is half of 90° (45°). Extend the lines for the 45° and ask him if this changes the size of the angle. It does not, since the angle depends on the amount of turning, not the length of the lines.

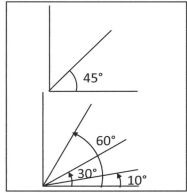

Draw another 90 degree angle and divide it into thirds (approximately). Ask your student for the size of an angle that is one third of 90° (30°) and two thirds of 90° (60°).

Divide the 30° angle into thirds and mark the 10° angle. Tell your student she should be able to recognize angles that are approximately 10°, 30°, 45°, and 60°, as well as 90°, 180°, and 270°. She can then refer to these angles for comparison in estimating other angles or making rough sketches. For example, an angle that is about a third of the way past 90° would be about 120°.

Draw some angles less than 180° and review measuring them with a protractor. For each, have your student estimate the angle first. Remind him that there are two curves on the protractor, each labeled with the degrees, and each starting at 0° on different sides of the protractor. Only the outer curve has markings between tens, but these markings can be used even when using the inner curve to find the degrees. Point out that the vertex of the angle must be placed at the center of the protractor's base line and the base line must be lined up along one of the arms of the angle.

Discussion

Concept p. 76

You can see if your student remembers how to measure angles greater than 180° by drawing one and asking her to measure it before looking at this page. Or you can immediately refer to the page in the textbook and discuss the two methods.

We can measure the difference from a half-turn, and then add that to 180°.

We can measure the opposite smaller angle, and then subtract that from 360°.

Practice

Task 1, p. 77

These may be easier to measure if they are traced and the sides extended. Note that measuring with a protractor is not absolutely accurate, especially if the angles are traced. Do not be concerned if your student's measures are a few degrees off from the answers given in this guide.

> 1. $\angle x = 124°$
> $\angle y = 239°$
> $\angle z = 325°$

Activity

Have your student practice drawing some angles of a given size. She will need to be able to draw angles accurately for Chapters 5 and 8 of this unit. Start with angles less than 180°. To draw an angle of a given measure, first draw a line with the bottom edge of the protractor and mark where the vertex of the angle will be. Line up the vertex point and the drawn line with the center of the protractor's base line. Mark a dot at the appropriate angle. Use the edge of the protractor to connect the vertex with the dot. For angles greater then 180° calculate the difference with 180° or 360° first.

Workbook

Exercise 1, pp. 69-72 (answers p. 97)

(2) Relate angles to direction and compass points

Activity

Show your student a compass and how to find north. Have him face north and tell you which direction is east, south and west.

Give your student a paper circle (appendix p. a9) and have her fold it into eighths by folding it in half and then half again twice, then unfold. Have her mark the creases and label the ends of the lines around the circle with the eight points of the compass, first north, south, east and west and then north-east, south-east, south-west, and north-west. She can use abbreviations.

You may want to discuss the use of these terms in everyday language. For example, show your student a map and have him tell you what areas on the map are in the south-west of a country, or what the direction from one point to another is.

Ask your student to find an angle between each line (45°). He could measure it, but he should be able to just tell you what it is. Discuss the meaning of turning in a clockwise or counterclockwise direction.

Discuss the angles between various points of the compass. For example, ask your student for the angle between north and southwest, both in a clockwise direction and a counterclockwise direction. The angles should be easily determined as being 90°, 180°, 270°, or 45° more than one of those.

Have your student stand up and face a given direction and then turn to another compass point and tell you how far he turned. For example, have him face north and turn clockwise to the south-east. He has turned 135°.

Ask your student to face a given direction and then turn 45°, 90°, 135°, 180°, 225°, 270°, or 315° clockwise or counterclockwise and tell you what direction she is facing. For example, ask her to face south-west, turn 270° clockwise, and give the direction she is now facing (south-east).

Practice

Tasks 2-3, p. 77

Workbook

Exercise 2, pp. 73-74 (answers p. 97)

Reinforcement

Extra Practice, Unit 9, Exercise 1, pp. 211-212

Test

Tests, Unit 9, 1A and 1B, pp. 97-102

2. (a) 135°
   (b) 45°

3. (a) north-east
   (b) north-west

Chapter 2 – Finding Unknown Angles

Objectives

♦ Recognize some angle properties involving intersecting lines.

♦ Find unknown angles using angle properties of intersecting lines.

Material

♦ Protractor

Notes

In this chapter your student will learn some angle properties that can be used to find unknown angles involving intersecting lines.

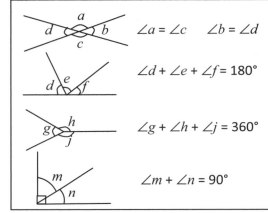

⇒ Vertically opposite angles are equal. (vert. opp. ∠'s)

$\angle a = \angle c$ $\angle b = \angle d$

⇒ The sum of the angles on a straight line is 180°. (∠'s on st. line)

$\angle d + \angle e + \angle f = 180°$

⇒ The sum of the angles at a point is 360°. (∠'s at a point)

$\angle g + \angle h + \angle j = 360°$

⇒ the sum of the angles formed by lines intersecting with a right angle is 90°. (∠'s in rt. ∠)

$\angle m + \angle n = 90°$

Your student will then use these properties to find unknown angles in a figure, given some known angles.

Angles that add up to 90° are called complementary angles and those that add up to 180° are called supplementary angles. It is not a requirement of this curriculum that students memorize those terms at this time.

As your student works through the tasks in the textbook, ask him to give you the reasons for each step in finding the unknown angle orally. Do not require him to write them down. In the practices and workbook problems you can ask him to write any equations he used.

Reasons are given in the solutions in this guide for your benefit, using the abbreviations given above. One suggested solution is given; there can be more than one approach for solving the problem.

Figures in the textbook for problems where students are to find unknown angles by applying the properties they learn are not drawn to scale. Your student cannot measure the angles to find the correct answer.

(1) Determine angle properties for intersecting lines

Discussion

Concept pp. 78-79

Your student should measure the angles for the diagrams on these pages. If they are hard to measure you can trace them and extend the lines. The angles have to be measured carefully.

After finishing p. 79, refer back to the intersecting lines on p. 78. Have your student add the angle measurements to find $\angle a + \angle b$, $\angle b + \angle c$, $\angle c + \angle d$, and $\angle d + \angle a$. Each pair adds to 180°.

You can have your student see if the properties on this page work in other cases by drawing some intersecting lines, measuring all the angles, comparing the vertically opposite ones, and finding the sum of the two along one of the straight lines. You can also draw other angles along a straight line or around a point for your student to measure the angles. You can use dynamic geometry software, such as Geometers Sketchpad®, or look for internet sites with embedded activities. Measuring to see if the properties hold true for other instances is rather tedious, and even using some dynamic geometry software program does not prove they really are true for all cases. More than likely the properties are self-evident to your student. Two of the properties follow from the definition of angles and the unit of measurement, and the third can be proven mathematically. Discuss how the properties listed on these two pages make sense.

The sum of the angle on a straight line is 180°: A straight line is a half-turn around the circle, or 180°. If the half-turn is "cut" into smaller turns, or angles, all the parts will add to 180°. Your student already saw this with turning to face different compass points.

The sum of the angles at a point is 360°: A full turn around a circle at a point is 360°. If the full turn is "cut" into smaller parts, all the parts will add to 360°.

$\angle a = 34°$	
$\angle b = \mathbf{146°}$	
$\angle c = \mathbf{34°}$	
$\angle d = \mathbf{146°}$	
$\angle q = \mathbf{85°}$	
$\angle r = \mathbf{45°}$	
$\angle p + \angle q + \angle r = \mathbf{180°}$	
$\angle y = \mathbf{150°}$	
$\angle z = \mathbf{150°}$	
$\angle x + \angle y + \angle z = \mathbf{360°}$	

Vertically opposing angles are equal: This can be proven from the property that the sum of the angle on a straight line is 180°, as shown at the right.

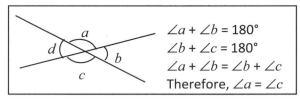

$\angle a + \angle b = 180°$
$\angle b + \angle c = 180°$
$\angle a + \angle b = \angle b + \angle c$
Therefore, $\angle a = \angle c$

Tasks 1-3, p. 80

Tell your student that these figures are not drawn exactly to scale, so she cannot find the unknown angles by direct measurement. She must use angle properties to find the unknown angles through calculation.

Remind your student that a little square drawn in an angle indicates that the angle is a right angle.

1. (a) $\angle p = \mathbf{48°}$ ($\angle$'s in rt. $\angle$)
   (b) $\angle q = \mathbf{143°}$ ($\angle$'s on st. line)
   (c) $\angle r = \mathbf{345°}$ ($\angle$'s at a point)

2. $\angle x = 180° - 46° = \mathbf{134°}$ ($\angle$'s on st. line)
   $\angle y = \angle w = \mathbf{46°}$ (vert. opp. $\angle$'s)
   $\angle z = \angle x = \mathbf{134°}$ (vert. opp. $\angle$'s)

3. $\angle COB = \angle AOD$ (vert. opp. $\angle$'s)
   $\angle AOD = \angle AOE + \angle EOD = 105° + 50° = 155°$
   $\angle COB = \mathbf{155°}$

Practice

Tasks 4-7, p. 81

For Tasks 6 and 7, tell your student to assume that the lines that look straight are straight. There are two straight lines in the figure in Task 6, and one in the first and the last figures in Task 7.

Workbook

Exercise 3, pp. 75-76 (answers p. 97)

Reinforcement

Extra Practice, Unit 10, Exercise 2, pp. 213-214

Test

Tests, Unit 10, 2A and 2B, pp. 103-106

---

4. $\angle DBE = \mathbf{90°}$ ($\angle$'s on st. line)

5. $\angle x = \mathbf{35°}$ ($\angle$'s at a point)

6. $\angle m = 180° - 90° - 50° = \mathbf{40°}$ ($\angle$'s on st. line)
   $\angle n = \angle m = \mathbf{40°}$ (vert. opp. $\angle$'s)

7. $\angle a = 180° - 75° - 76° = \mathbf{29°}$ ($\angle$'s on st. line)
   $\angle b = 360° - 60° - 90° - 65° = \mathbf{145°}$ ($\angle$'s at a point)
   $\angle c = 125° - 40° = \mathbf{85°}$ (vert. opp. $\angle$'s)

Chapter 3 – Sum of Angles of a Triangle

Objectives

♦ Recognize some angle properties involving triangles.
♦ Find unknown angles using angle properties of triangles.

Material

♦ Protractor
♦ Appendix pp. a10-a11

Notes

In this chapter your student will learn some angle properties of triangles

⇒ The sum of the angles of a triangle is 180°.
   (∠ sum of Δ)

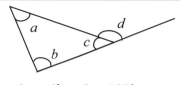

$\angle a + \angle b + \angle c = 180°$
If $\angle b$ is a right angle, then
$\angle a + \angle c = 90°$
$\angle d = \angle a + \angle b$

⇒ When one angle of a triangle is a right angle, the sum of the other two angles is 90°.
   (right Δ)

   Conversely, if the sum of two of the angles is not 90°, then the third angle is not 90° and the triangle is not a right triangle.

⇒ The exterior angles of a triangle are equal to the sum of the interior opposite angles.
   (ext. ∠ of Δ)

Your student will then use these properties to find unknown angles in a triangle, given some known angles.

In the textbook, right triangles are called right-angled triangles.

(1) Find the sum of angles of a triangle

Discussion

Concept p. 82

You may want to have your student do this activity before looking at the textbook page. You can have him try it with a variety of triangles.

Tasks 1-2, p. 83

Task 1: You can use the larger triangles on appendix p. a10; they are larger so the angles will be easier to measure. They have to be measured carefully if the sum is to be 180° for each.

Task 2: Since the sum of the angles is 180°, if we know two of the angles we can find the third by subtracting the other two from 180°.

> 1. A: 90° + 53° + 37° = 180°
>   B: 76° + 65° + 39° = 180°
>   C: 123° + 33° + 24° = 180°
>
> 2. ∠BCA = **44°**

Practice

Task 3, p. 83

> 3. $\angle a = 180° - 60° - 52° = \mathbf{68°}$
> $\angle b = 180° - 40° - 38° = \mathbf{102°}$
> $\angle c = 180° - 65° - 72° = \mathbf{43°}$

Discussion

Tasks 4-5 p. 84

Task 4: You may want to have your student do this activity before looking at the textbook page. She can try it with a variety of right triangles by simply cutting off the corners of a piece of paper with a straight line at different angles to get different right triangles.

The property that the sum of the two non-right angles in a right triangle is 90° can easily be derived from the property that all the angles of a triangle add up to 180°. If one angle is a right triangle, the sum of the other two must be 180° − 90° = 90°.

> 4. The two other triangles fit in the third right angle of the triangle.
>
> 5. ∠PRQ = **33°**

Practice

Task 6, p. 84

Workbook

Exercises 4-5, pp. 77-78 (answers p. 97)

> 6. A: 65° + 25° = 90°
>   B: 48° + 43° = 91°
>   C: 53° + 37° = 90°
>   A and C are right-angled triangles.

(2) Find unknown angles in triangles

Discussion

Task 7, p. 85

You can have your student do this activity before looking at the textbook. Draw a triangle, cut it out, then trace around it on paper. Extend one side of the traced triangle. Have him tear off the opposite angles of the cut-out triangle and fit them in the exterior angle. Then have him look at the textbook.

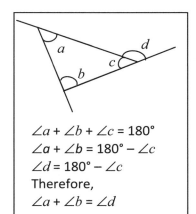

$\angle a + \angle b + \angle c = 180°$
$\angle a + \angle b = 180° - \angle c$
$\angle d = 180° - \angle c$
Therefore,
$\angle a + \angle b = \angle d$

See if your student can derive the property that the exterior angle of a triangle is equal to the sum of the interior opposite angles from the angle properties she already knows. Since the sum of the interior angles is 180°, then the sum of any two of the angles is the difference between 180° and the third angle. However, since the third angle and the exterior angle are on a straight line, the exterior angle is also the difference between 180° and the third angle. Therefore, the exterior angle is equal to the sum of the opposite interior angles.

Practice

Tasks 8-9, p. 85

8. $\angle$XZP = **84°**

9. $\angle a = 90° + 43° = $ **133°**
   $\angle b = 110° - 50° = $ **60°**

Activity

Give your student a copy of appendix p. a11. These are problems that require more than one step. Tell him that if he can't see right away how to solve this kind of problem, he should just start filling in any angle measurements he can determine. There are usually several different approaches. For example, in the first problem he can find the angle $\angle$ACB since it is on a straight line next to the 72° angle. Or, he could remember the property that an exterior angle of a triangle is the sum of the opposite interior angles and find $\angle$CAB first.

Appendix p. a11 answers

1. $\angle$CAB = 72° − 35° = 37°   (ext. $\angle$ of $\triangle$)
   $\angle a$ = 180° − 37° = 143°   ($\angle$'s on st. line)

2. $\angle$FHE = 48°   (vert. opp. $\angle$'s)
   $\angle b + \angle b$ = 180° − 48°
          = 132°   ($\angle$ sum of $\triangle$)
   $\angle b$ = 132° ÷ 2 = 66°

3. $\angle$NLM = 90° − 32° = 58°   ($\angle$'s in rt. $\angle$)
   $\angle c$ = 90° − 58° = 32°   (right $\triangle$)

Workbook

Exercise 6, p. 79 (answers p. 97)

Reinforcement

Extra Practice, Unit 10, Exercise 3, pp. 215-218

Test

Tests, Unit 10, 3A and 3B, pp. 107-110

Chapter 4 – Isosceles and Equilateral Triangles

Objectives

♦ Recognize some angle properties involving isosceles and equilateral triangles.
♦ Find unknown angles using angle properties of intersecting lines and triangles.

Material

♦ Protractor
♦ Straws

Vocabulary

♦ Isosceles triangle
♦ Equilateral triangle
♦ Scalene triangle

Notes

In *Primary Mathematics* 3B students learned to classify triangles based on the length of the sides. This was reviewed in *Primary Mathematics* 4A and students looked at symmetry properties of isosceles and equilateral triangles.

An **isosceles** triangle has two equal sides.

An **equilateral triangle** has three equal sides.

A **scalene triangle** has all sides different in length.

Equal sides on a geometric figure are marked with a small cross-hatch.

In this chapter students will learn about angle properties of the three different types of triangles.

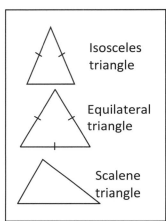

⇒ In an isosceles triangle, the two angles facing the equal sides (the "base" angles) are equal.
(iso. Δ)

Conversely: If two angles of a triangle are the same, then the triangle is an isosceles triangle.

⇒ In an equilateral triangle, all three angles equal 60°.
(equ. Δ)

Conversely: If each of the three angles of a triangle is 60°, then the triangle is an equilateral triangle.

If we are told that two of the angles are 60°, then we know it is an equilateral triangle, because the third angle must be 60°.

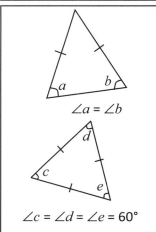

An equilateral triangle is a special type of isosceles triangle.

Your student will be finding unknown angles using all the angle properties she has learned. Do not require that she write down the reason for each step of the calculation for finding the unknown angle at this level. Encourage her to give the reasons during oral discussion. There can be more than one solution to many of the problems.

(1) Learn angle properties of isosceles and equilateral triangles

Discussion

Concept p. 86

Before looking at the textbook, you may want to have your student discover the angle properties for isosceles and equilateral triangles himself. Use straws or thin cardboard strips and have him cut 3 pieces, two of which are equal, use them to form a triangle, trace it, and measure the angles. Then cut the other two sides to be the same length of the third and repeat. Within experimental error, he should find that two of the angles are the same, and if all three sides are the same length, all three angles are the same.

Remind your student that a triangle with no equal sides is called a scalene triangle. You can explain the meaning of the word parts for isosceles, equilateral, and scalene. It is easy to mix up the words isosceles and equilateral.

Tasks 1-4, pp. 87-88

Task 1: Remind your student that isosceles triangles have a line of symmetry. You can have her draw an isosceles triangle, cut it out, and fold it along the line of symmetry. The two base angles will overlap exactly. To draw an isosceles triangle, all she has to do is draw two straight lines of equal lengths that form an angle and then join the ends with a straight line.

Task 4: Since an equilateral triangle has three lines of symmetry, each pair of angles have to be the same, so all three have to be the same. $180° \div 3 = 60°$

Workbook

Exercise 7, pp. 80-81 (answers p. 97)

A, B, C, and D have 2 equal sides. They are called isosceles triangles.
A and C have 3 equal sides and are called equilateral triangles.

---

equi-: equal   lateral: sides
Equilateral triangle: triangle with equal sides

Iso-: equal -skelos: legs
Isosceles triangle: triangle with "equal legs"

Scalenus: unequal
Scalene triangle: triangle with unequal sides

---

1. See textbook.

2. See textbook.

3. ΔA is not isosceles. The third angle will be greater than 90°.
   ΔB is isosceles. Two of the angles are equal.
   ΔC is isosceles. $180° - 75° - 30° = 75°$.

4. **ΔP** is equilateral. The third angle must be 60°.
   ΔQ is equilateral. Two sides are equal so another angle is 60°.
   ΔR is not equilateral. One angle is not 60°.

(2) Find unknown angles

Discussion

Tasks 5-7, pp. 88-89

Task 5: We know ∠ACB because it is a base angle in an isosceles triangle and we are given the measure of the other base angle.

Point out that if we are given any angle in an isosceles triangle we know the other two angles. Draw an isosceles triangle, label the non-base angle with a measurement, such as 65°, and ask your student to find the other two angles. Since it is an isosceles triangle, the other two angles are equal. So she just divides the difference between the first angle and 180° by 2.

Draw an isosceles right triangle and ask your student to find the other two angles.

Task 6: Tell your student to always pay attention to the cross-hatches to know which two sides and so which two angles are equal. Ask him to find ∠QRP as well.

Task 7: Ask your student which angles are external angles of the triangle (∠ACD and ∠BCE). Ask him to find ∠BAC as well.

Practice

Task 8, p. 89

Workbook

Exercise 8, pp. 82-83 (answers p. 97)

Reinforcement

Extra Practice, Unit 10, Exercise 4, pp. 219-222

Test

Tests, Unit 10, 4A and 4B, pp. 111--116

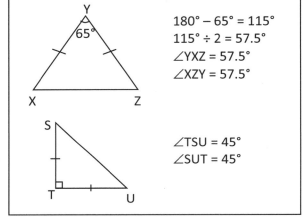

5. ∠ACB = 35°   (iso. Δ)
   ∠ACB = 180° − 35° − 35° = **110°** (∠ sum of a Δ)

6. ∠QPR = 65°   (iso. Δ)
   ∠PRS = 65° + 65° = **130°**   (ext. ∠ of a Δ)
   ∠QRP = 180° − 130° = 50°

7. ∠ACB = ∠DCE = 75°   (vert. opp. ∠'s)
   ∠ABC = ∠ACB = **75°**   (iso. Δ)
   ∠BAC = 180° − (2 x 75°) = 30°

180° − 65° = 115°
115° ÷ 2 = 57.5°
∠YXZ = 57.5°
∠XZY = 57.5°

∠TSU = 45°
∠SUT = 45°

8. ∠ACB = 50°   (iso. Δ)
   ∠a = 180° − 50° = **130°**   (∠'s on a st. line)

   ∠ACB = 180° − 110° = 70°   (∠'s on a st. line)
   ∠b = 180° − (2 x 70°) = **40°**   (iso. Δ)

   ∠BAC = ∠ABC = 60°   (equ. Δ)
   ∠c = 60° + 60° = **120°**   (ext. ∠ of a Δ)

   ∠BCA = 60°   (equ. Δ)
   ∠d = 180° − 60° − 90° = **30°**   (∠'s on a st. line)

   ∠BAC = ∠ABC = 60°   (equ. Δ)
   ∠ACD = 60° + 60° = 120°   (ext. ∠ of a Δ)
   ∠e = 180° − 40° − 120° = **20°**   (∠ sum of a Δ)

Workbook

Exercise 1, pp. 69-72

1. $\angle a = 38°$ $\angle b = 65°$
   $\angle c = 90°$ $\angle d = 121°$
   $\angle e = 160°$ $\angle f = 180°$
   $\angle g = 202°$ $\angle h = 245°$
   $\angle i = 270°$ $\angle j = 307°$
   $\angle k = 338°$ $\angle l = 360°$

2. $\angle a = 67°$ $\angle b = 230°$ $\angle c = 129°$ $\angle d = 335°$

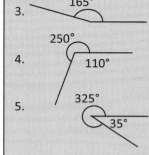

3. 165°

4. 250°   110°

5. 325°   35°

Exercise 2, pp. 73-74

1. south
   north-east
   south-east
   west
   north-west
   east
   south-west

2. north-east
   east
   south-east
   north-west
   45°
   90°
   135°
   45°

Exercise 3, pp. 75-76

1. $\angle a = 180° - 135° = \textbf{45°}$ ($\angle$'s on st. line)
   $\angle b = 90° - 32° = \textbf{58°}$ ($\angle$'s in a rt. $\angle$)
   $\angle c = \textbf{48°}$ (vert. opp. $\angle$'s)
   $\angle d = 360° - 24° = \textbf{336°}$ ($\angle$'s at a point)
   $\angle e = 360° - 250° = \textbf{110°}$ ($\angle$'s at a point)
   $\angle f = 180° - 90° = \textbf{90°}$ ($\angle$'s on st. line)
   $\angle g = 180° - 28° = \textbf{152°}$ ($\angle$'s on st. line)
   $\angle h = \textbf{145°}$ (vert. opp. $\angle$'s)

2. $\angle q = 180° - 72° - 80° = \textbf{28°}$ ($\angle$'s on st. line)
   $\angle r = 180° - 35° - 28° = \textbf{117°}$ ($\angle$'s on st. line
   $\angle s = 360° - 108° - 85° - 90° = \textbf{77°}$
                            ($\angle$'s at a point)
   $\angle t = 360° - 135° - 124° = \textbf{101°}$ ($\angle$'s at a point)
   $\angle u = 360° - 92° - 78° - 140° = \textbf{50°}$
                            ($\angle$'s at a point)

Exercise 4, p. 77

1. (a) $\angle ACB = 180° - 76° - 72° = \textbf{32°}$
   (b) $\angle TRS = 180° - 90° - 48° = \textbf{42°}$
   (c) $\angle LMK = 180° - 28° - 28° = \textbf{124°}$
   (d) $\angle FGH = 180° - 118° - 42° = \textbf{20°}$

Exercise 5, p. 78

1. (a) $\angle a = 90° - 15° = \textbf{75°}$
   (b) $\angle b = 180° - 85° - 56° = \textbf{39°}$
   (c) $\angle c = 180° - 45° - 45° = \textbf{90°}$
   (d) $\angle d = 180° - 60° - 35° = \textbf{85°}$
   (a) and **(c)** are right-angled triangles.

Exercise 6, p. 79

1. (a) $\angle ABD = 65° + 58° = \textbf{123°}$
   (b) $\angle WXZ = 90° + 46° = \textbf{136°}$
   (c) $\angle BDC = 125° - 86° = \textbf{39°}$
   (d) $\angle SRT = 56° - 25° = \textbf{31°}$

Exercise 7, pp. 80-81

1. (a) $\angle a = 180° - 64° - 73° = \textbf{43°}$
   (b) $\angle b = 180° - 76° - 52° = \textbf{52°}$
   (c) $\angle c = 180° - 35° - 35° = \textbf{110°}$
   (d) $\angle a = 180° - 90° - 47° = \textbf{43°}$
   (b) and **(c)** are isosceles triangles.

2. (a) $a = \textbf{60°}$
   (b) $b = 90° - 60° = \textbf{30°}$
   (c) $c = 180° - 50° - 65° = \textbf{65°}$
   (d) $d = \textbf{60°}$
   (a) and **(d)** are equilateral triangles.

Exercise 8, pp. 82-83

1. (a) $\angle a = 180° - 55° - 55° = \textbf{70°}$ ($\angle$ sum of Δ)
   (b) $\angle b = 90° - 37° = \textbf{53°}$ (right Δ)
   (c) $\angle c = 110° + 35° = \textbf{145°}$ (ext. $\angle$ of Δ)
   (d) $\angle d = 180° - 107° - 45° = \textbf{28°}$ ($\angle$ sum of Δ)
   (e) $\angle ACB = 180° - 118° = 62°$ ($\angle$'s on st. line)
       $\angle e = \angle ACB = \textbf{62°}$ (iso. Δ)
   (f) $\angle DBC = 60°$ (equ. Δ)
       $\angle f = 180° - 60° = \textbf{120°}$ ($\angle$'s on st. line)
   (g) $\angle ACB = 60°$ (equ. Δ)
       $\angle g = 180° - 60° - 70° = \textbf{50°}$ ($\angle$'s on st. line)
   (h) $\angle h = 135° - 90° = \textbf{45°}$ (ext. $\angle$ of a Δ)

Chapter 5 – Drawing Triangles

Objectives

◆ Construct a triangle, given the measurements of two angles and their included side.
◆ Construct a triangle, given the measurements of two sides and their included angle.

Material

◆ Protractor
◆ Ruler
◆ Set square
◆ Appendix p. a12

Notes

In *Primary Mathematics* 4A students learned to draw perpendicular and parallel lines using a ruler and a set-square (a plastic triangle with one 90 degree angle, see p. 99 of the textbook for a picture of a set-square). They also learned how to estimate and draw an angle of a given size using a protractor.

In this chapter your student will learn to draw triangles given the measurements of two angles and their included side or two sides and their included angle.

When a picture is not included for reference, your student should make a quick sketch of the desired triangle estimating the angles before making a full-size drawing with exact measurements in order to get an idea of what the final drawing should look like.

(1) Draw triangles

Discussion

Concept p. 90
Tasks 1-2, pp. 91-92

Guide your student through actually drawing these triangles, not just looking at the directions.

Practice

Task 3, p. 92

Task 3(a): Ask your student to measure length BC and angle ACB.

Tasks 3(b)-3(d): Ask your student to find the length of AC.

> 3. (a) Draw AB = 7 cm.
>   Draw a line perpendicular to AB through A, using a set-square.
>   Mark AC = 5 cm.
>   Join BC.
>   BC ≈ 8.6 cm; ∠ACB ≈ 54.5
> (b) Draw AB = 8 cm.
>   Draw ∠BAC = 40° and ∠ABC = 60°.
>   Mark the intersection of these lines as C.
>   AC ≈ 7 cm
> (c) Draw AB = 6 in.
>   Draw ∠ABC = 50° such that BC = 5 in.
>   Join AC.
>   AC ≈ 4.75 in.
> (d) Draw AB = 7 in.
>   Draw ∠BAC = 30° and ∠ABC = 100°.
>   Mark the intersection of these lines as C.
>   AC = 9 in.

Activity

Give your student a copy of appendix p. a12 or write the directions on a whiteboard or paper. Have her first draw a small sketch of what the triangle should look like before attempting to construct the full-sized triangle. A scaled drawing and answers are given at the right.

Workbook

Exercise 9, p. 84 (answers p. 107)

Reinforcement

Extra Practice, Unit 10, Exercise 5, pp. 223-224

Test

Tests, Unit 10, 5A, pp. 117-118

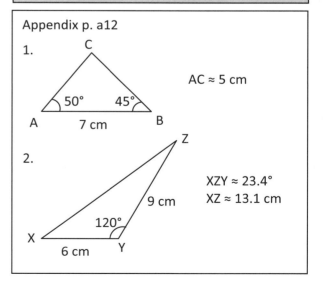

Appendix p. a12

1.

AC ≈ 5 cm

2.

XZY ≈ 23.4°
XZ ≈ 13.1 cm

Chapter 6 – Sum of Angles of a Quadrilateral

Objectives

♦ Recognize some angle properties involving quadrilaterals.
♦ Find unknown angles using angle properties of quadrilaterals.

Vocabulary

Quadrilateral

Notes

In this chapter your student will find the sum of the angles of any quadrilateral:

⇒ The sum of the angles of a quadrilateral is 360°.
  (∠ sum of ◊)

Any quadrilateral (4-sided figure) can be divided into two triangles. The sum of the angles of these two triangles together is the same as the sum of the angles of the quadrilateral. The total sum of the angles of the quadrilateral is therefore 360°.

Your student will use this property to find an unknown angle in a quadrilateral, given the other three angles.

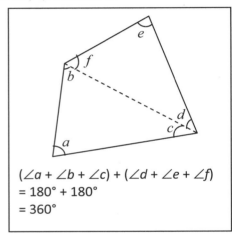

$(\angle a + \angle b + \angle c) + (\angle d + \angle e + \angle f)$
$= 180° + 180°$
$= 360°$

For your information, a polygon of n sides can be divided up into $n - 2$ triangles, so the sum of the angles is 180° x $(n - 2)$. If the polygon is a regular polygon (equilateral and equiangular) then the measure of each angle is $\dfrac{180 \times (n-2)}{n}$.

(1) Find the sum of angles of a quadrilateral

Discussion

Concept p. 93

You may want to have your student do this activity before looking at the textbook page. He can draw any kind of quadrilateral, and try a variety of quadrilaterals. Remind him that a quadrilateral is a closed figure with 4 sides.

Task 1, p. 94

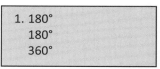

This task shows a method to prove that the sum of angles in a quadrilateral is 360° for all quadrilaterals, if one accepts that the sum of angles in any triangle is 180°. Any quadrilateral can be divided into two triangles. You can show that this is true even for convex quadrilaterals.

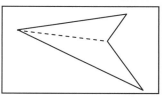

Practice

Task 2, p. 94

Workbook

Exercise 10, p. 85 (answers p. 107)

Reinforcement

Extra Practice, Unit 10, Exercise 6, pp. 225-226

Test

Tests, Unit 10, 6A and 6B, pp. 119-122

Note: the problems in Test 6A are somewhat harder than any in the textbook and workbook. This test could be used for enrichment rather than as a test. If your student has difficulties with any of the problems, tell her to start filling in the measurements for any angles she can find until she sees a way to find the required angle. These types of problems are good for developing problem-solving skills based on deductive reasoning and often have more than one approach.

Enrichment

Have your student use the proof in Task 1 to find the sum of angles in a pentagon, hexagon, and octagon, and see if he can come up with a general rule for the sum of the angles in a closed figure of any number of sides.

Chapter 7 – Parallelograms, Rhombuses and Trapezoids

Objectives

♦ Recognize some angle properties involving parallelograms and trapezoids.
♦ Find unknown angles using angle properties of parallelograms, trapezoids, and triangles.

Material

♦ Appendix p. a13

Vocabulary

♦ Parallelogram
♦ Rhombus
♦ Trapezoid

Notes

In *Primary Mathematics* 4A students learned to classify quadrilaterals based on the number of parallel sides, right angles, and equal sides. In this chapter your student will review the names of quadrilaterals with at least one pair of parallel sides and learn some angle properties for parallelograms and trapezoids.

The diagram below gives the definition of specific types of quadrilaterals your student has learned about and their relationship to each other.

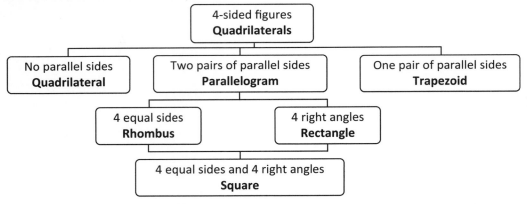

Two angle properties of parallelograms are introduced in this section.

⇒ The opposite angles of a parallelogram are equal.
   (opp. ∠'s of //)

⇒ Each pair of angles between two parallel sides adds up to 180°.
   (inside ∠'s //)

Since rectangles, rhombuses, and squares are also parallelograms, these properties apply to them as well.

Trapezoids have one set of parallel sides, so the pair of angles between the parallel sides add up to 180°.

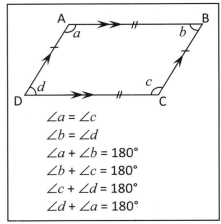

$\angle a = \angle c$
$\angle b = \angle d$
$\angle a + \angle b = 180°$
$\angle b + \angle c = 180°$
$\angle c + \angle d = 180°$
$\angle d + \angle a = 180°$

In drawings an equal number of markings indicate equal lines (marked by short cross hatches) or parallel lines (marked by arrows going in the same direction). In the diagram above, the cross hatches and arrows indicate that AB = DC, DA = CB, AB // DC, and AD // BC.

(1) Find angle properties of parallelograms

Discussion

Concept p. 95

This activity is meant to review the properties of different types of quadrilaterals. Give your student a copy of appendix p. a13 so that she can write on the paper and mark the figures.

Ask your student to identify the figures with two pairs of parallel lines (A, B, D, and E). Ask her what they are called (parallelograms), and have her label them. Show her how she can indicate the lines that are parallel with small arrows. Sides with the same number of arrowheads are parallel. There is no significance to the direction of the arrowheads or which side gets the double ones.

Then have your student identify the figures with one pair of parallel lines (C and F), name and label them (trapezoids), and mark parallel sides.

Then have your student identify the figures with equal sides (A and E). Remind him, if necessary, that parallelograms with equal sides are called rhombuses and have him mark equal sides and label each as a rhombus. Only one mark is needed on each side. He can also mark which sides are equal on the other parallelograms.

Then have your student identify the figures with right angles (B and E), name them (both are rectangles, E is also a square) and mark the right angles. Point out that all squares have the same properties as rectangles, rhombuses, and parallelograms. All rectangles and rhombuses have the same properties as parallelograms. But the converse is not true; all parallelograms do not have the same properties as rectangles.

Tasks 1-2, pp. 96-97

Your student needs two copies of the parallelogram (the cuts for 1 and 2(a) are the same). She can trace the one in the textbook, or cut out the ones on appendix p. a13, or both. Before cutting out the parallelograms, have her label the corners. Point out that these two properties are obviously true for rectangles and squares as well.

Practice

Task 3, p. 97

3. 80°	60 °	65°

Workbook

Exercise 11, pp. 86-87 (answers p. 107)

Note: If a problem in the workbook does not specifically say the figure is a rhombus, rectangle, or square, do not assume it is by visual inspection.

(2) Find unknown angles

Discussion

Tasks 4-6, p. 98

Task 5: Since the sides of a rhombus are equal, each diagonal line cuts the rhombus into two isosceles triangles. Your student probably will remember that a rhombus has two lines of symmetry. Problems involving finding unknown angles in rhombuses in the textbook or workbook often require use of angle properties of triangles as well as parallelograms.

Task 6: Whenever a line traverses two parallel lines, the sum of the pair of angles between the two parallel lines on one side of the transversal is 180°. You can illustrate this by drawing two parallel lines and then drawing some parallelograms and trapezoids using transverse lines.

It is important to pay attention to which lines are parallel. In this task, the sum of angles at D and C and at A and B are 180°. The sum of the angles at A and D or B and C are not.

4. $\angle$YXZ = 90° − 26° = **64°**   ($\angle$'s in a rt. $\angle$)

5. $\angle x$ = **60°**   (opp. $\angle$'s of //)
   $\angle y$ = 180° − 130° = **50°**   (inside $\angle$'s //)
   $\angle z$ = **55°**   (iso. Δ)
   The sides of a rhombus are equal, so $\angle z$ is a base angle of an isosceles triangle.

6. $\angle$ABC = **130°**
   $\angle$DCB = **60°**

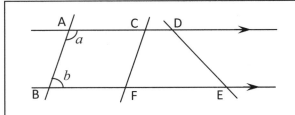

ACFB is a parallelogram.
$\angle a$ + $\angle b$ = 180°
ADEB is a trapezoid.

Practice

Task 7, p. 98

7. $\angle a$ = 180° − 125° = **55°**   (inside $\angle$'s //)

   $\angle b$ = 180° − 68° = **112°**   (inside $\angle$'s //)

   $\angle$ next to 22° angle = 50°   (iso. Δ)
   $\angle c$ = 180° − (22° + 50°) = **108°** (inside $\angle$'s //)

Workbook

Exercise 12, pp. 88-90 (answers p. 107)

Reinforcement

Extra Practice, Unit 10, Exercise 7, pp. 227-230

Test

Tests, Unit 10, 7A and 7B, pp. 123-128

Chapter 8 – Drawing Parallelograms

Objectives

♦ Construct a parallelogram.
♦ Construct a rectangle with a given length and width.
♦ Construct a parallelogram when given the measurement of two adjacent sides and one angle.
♦ Construct a rhombus when given the measurement of one side and one angle.

Material

♦ Protractor
♦ Ruler
♦ Set square

Notes

In *Primary Mathematics* 4A students learned to draw perpendicular and parallel lines using a ruler and a set-square (a plastic triangle with one 90 degree angle, see p. 99 of the textbook for a picture of a set-square). They also learned how to estimate and draw an angle of a given size using a protractor. In this chapter your student will learn to draw parallelograms and rhombuses from given measurements of some of the sides and angles.

If a sketch isn't provided in the problem, your student should make a rough sketch of the figure before drawing it full scale.

Although parallelograms and rhombuses can be drawn using compasses and intersecting circles, students will not be learning those techniques at this level.

(1) Draw parallellograms

Discussion

Concept p. 99
Tasks 1-2, pp. 100-101

Have your student actually draw these parallelograms, not just look at the directions. The third side can be drawn using a set-square or ruler to draw a parallel line, or the same way as the second side by extending the line and using the same measurement for the outside angle, or using a measure for the inside angle that is the difference between the given angle and 180°. Parallelograms are actually easier to construct than triangles because opposite sides are parallel and have the same lengths.

Practice

Tasks 3-5, p. 102

> 3. Draw PQ = 6 cm.
>   Draw PS such that $\angle SPQ = 110°$ and PS = 5 cm.
>   Draw QR such that QR // PS and QR = 5 cm.
>   Join SR.
>
> 4. Draw AB = 4 cm.
>   Draw AD such that $\angle DAB = 40°$ and AD = 4 cm.
>   Draw BC such that BC // AD and BC = 4 cm.
>   Join DC.
>
> 5. Draw AB = 8 cm.
>   Draw BC such that $\angle ABC = 100°$ and BC = 8 cm.
>   Draw AD such that AD // BC and AD = 8 cm.
>   Join DC.

Workbook

Exercise 13, pp. 91-92 (answers p. 107)

Reinforcement

Extra Practice, Unit 10, Exercise 8, pp. 231-232

Test

Tests, Unit 10, 8A, pp. 129-130

Workbook

Exercise 9, p. 84

1. Draw AB = 5 cm.
   Draw ∠BAC = 56° and ∠ABC = 78°
   Mark the intersection of these lines as C.

2. Draw XY = 6 cm.
   Draw a line perpendicular to XY through Y, using
   a set-square or a protractor.
   Mark YZ = 4 cm.
   Join XZ.

Exercise 10, p. 85

1. (a) $\angle x = 360° - 125° - 75° - 75° = $ **85°**
   (b) $\angle x = 360° - 135° - 70° - 80° = $ **75°**
   (c) $\angle x = 360° - 88° - 75° - 84° = $ **113°**
   (d) $360° - 138° - 75° - 83° = 64°$
       $\angle x = 180° - 64° = $ **116°**   (∠'s on a st. line)
   (e) $360° - 155° - 48° - 62° = 95°$
       $\angle x = $ **95°**   (vert. opp. ∠'s)

Exercise 11, pp. 86-87

1. (a) $\angle a = $ **55°**   (opp. ∠'s of //)
   (b) $\angle b = 180° - 75° = $ **105°**   (inside ∠'s //)
   (c) $\angle c = $ **125°**   (opp. ∠'s of //)
   (d) $\angle d + 18° + 142° = 180°$ (inside ∠'s //)
       $\angle d = 180° - 18° - 142° = $ **20°**
   (e) $\angle e = $ **110°**   (opp. ∠'s of //)
   (f) $\angle f + 60° + 60° = 180°$ (inside ∠'s //)
       $\angle f = 180° - 60° - 60° = $ **60°**
   (g) $\angle g = 180° - 80° = $ **100°**   (inside ∠'s //)
   (h) $\angle h = $ **135°**   (opp. ∠'s of //)

Exercise 12, pp. 88-90

1. (a) $\angle CBD = 90° - 60° = $ **30°**   (∠'s in a rt. ∠)
   (b) $\angle PRQ = 90° - 55° = $ **35°**   (∠'s in a rt. ∠)

2. (a) $\angle a = $ **100°**   (opp. ∠'s of //)
   (b) $\angle b = 180° - (2 \times 75°) = $ **30°** (iso. Δ)
   (c) $\angle c = $ **56°**   (opp. ∠'s of //)
   (d) $\angle d = 180° - 73° = $ **107°**   (inside ∠'s //)
   (e) $\angle e = 180° - 40° = $ **140°**   (inside ∠'s //)
   (f) $\angle f + 50° + 80° = $ **180°**   (inside ∠'s //)
       $\angle f = 180° - 50° - 80° = $ **50°**

3. (a) $\angle a = 180° - 112° = $ **68°**   (inside ∠'s //)
   (b) $\angle b + 32° + 123° = 180°$ (inside ∠'s //)
       $\angle b = 180° - 32° - 123° = $ **25°**
   (c) $\angle c = 180° - 84° = $ **96°**   (inside ∠'s //)
       $\angle x = 180° - 48° = $ **132°**   (inside ∠'s //)
   (d) $\angle y + 87° + 25° = 180°$ (inside ∠'s //)
       $\angle y = 180° - 87° - 25° = $ **68°**
       $\angle d = 180° - \angle y$ (inside ∠'s //)
       $\angle d = 180° - 68° = $ **112°**   (inside ∠'s //)

Exercise 13, pp. 91-92

1. Draw BC such that ∠ABC = 120° and BC = 5 cm.
   Draw AD such that AD // BC and AD = 5 cm.
   Join DC.

2. Draw PQ = 7 cm.
   Draw PS such that ∠SPQ = 45° and PS = 6 cm.
   Draw QR such that QR // PS and QR = 6 cm.
   Join SR.

3. Draw AD such that ∠DAB = 50° and AD = 6 cm.
   Draw BC such that BC // AD and BC = 6 cm.
   Join DC.

4. Draw PQ = 5 cm.
   Draw PS such that ∠SPQ = 120° and PS = 5 cm.
   Draw QR such that QR // PS and QR = 5 cm.
   Join SR.

Review 10

Review

Review 10, pp. 103-105

Workbook

Review 10, pp. 93-96 (answers p. 109)

Tests

Tests, Units 1-10, Cumulative Tests A and B, pp. 131-137

---

1. (a) $16 + \underline{3 \times 8} \div 4$
   $= 16 + \underline{24 \div 4}$
   $= 16 + 6$
   $= \mathbf{22}$

   (b) $30 + \underline{85 \times 2} \div (8 + 9)$
   $= 30 + \underline{170 \div 17}$
   $= 30 + 10$
   $= \mathbf{40}$

   (c) $\underline{(220 \div 11)} \times (28 - 5)$
   $= 20 \times 23$
   $= \mathbf{460}$

   (d) $12 + \underline{(30 - 14)} \div 4 \times 5$
   $= 12 + \underline{16 \div 4} \times 5$
   $= 12 + \underline{4 \times 5}$
   $= 12 + 20$
   $= \mathbf{32}$

2. (a) 82   (b) 27   (c) 18

---

3. 2 units = 120
   1 unit = 120 ÷ 2 = 60
   Green: 1 unit = **60**
   There are 60 green paper clips.

4. $\frac{3}{4}$ kg ÷ 3 = $\frac{3}{4}$ kg × $\frac{1}{3}$ = $\frac{\mathbf{1}}{\mathbf{4}}$ **kg**

   Each bag weighs $\frac{1}{4}$ kg.

5. 5 units = $2500
   1 unit = $2500 ÷ 5 = $500
   3 units = $500 × 3 = **$1500**
   He spends $1500.

6. Total units = 8
   8 units = 96 cm
   1 unit = 96 ÷ 8 = 12 cm
   Longest rod = 4 units
   4 units = 12 × 4 = **48 cm**

7. 1 unit = $2.50
   (5 × $2.50) + $5.00
   = **$17.50**
   They have $17.50 altogether.

8. (20 m – 6.32 m) ÷ 8 = 13.68 m ÷ 8 = **1.71 m**
   Each piece is 1.71 m long.

9. (a) $\frac{6}{8}$ × 100% = **75%**   (b) $\frac{7}{10}$ × 100% = **70%**

10. 0.8

11. 36% = $\frac{36}{100}$ = $\frac{\mathbf{9}}{\mathbf{25}}$

12. (a) 90%   (b) 8%   (c) 58%   (d) 9%

13. (a) $\frac{7}{100}$ × 160
    = **11.2**
    (b) $\frac{80}{100}$ × 98 kg
    = **78.4 kg**
    (c) $\frac{15}{100}$ × $21
    = **$3.15**

---

14. 98% of 150 = $\frac{98}{100}$ × 150 = **147**
    147 of the students passed the test.

15. 100% − 90% = 10%
    10% of 250 = **25**
    There were 25 doughnuts left.

16. 4% of $5000 = $\frac{4}{100}$ × $5000 = 4 × $50 = $200
    $5000 + $200 = **$5200**
    She has $5200 in the bank.

17. Discount: 20% of $45 = $\frac{20}{100}$ × $45 = $9
    Selling price: $45 − $9 = **$36**
    The selling price is $36.

18. (a) Area of shaded part
       = area of rectangle − area of un-shaded part
       = (8 cm × 12 cm) − $\frac{1}{2}$ × (6 cm × 8 cm)
       = 96 − 24 = **72 cm²**
    (b) The 8 in. base has a 5 in. height.
       Area of triangle = $\frac{1}{2}$ × (8 in. × 5 in.) = **20 in.²**

19. (a) $\angle x = 360° − 240° − 90° = \mathbf{30°}$ (∠'s at a point)
    (b) $\angle COE + \angle EOB = \angle AOD$ (vert. opp. ∠'s)
       $\angle x = 110° − 90° = \mathbf{20°}$

20. (a) 161°   (b) 218°

21. (a) $x = 360° − 63° − 90° = \mathbf{207°}$ (∠'s at a point)
    (b) $180° − 90° − 43° = 47°$ (∠'s on st. line)
       $x = \mathbf{47°}$ (vert. opp. ∠'s)

22. (a) 7 cm × 5 cm = **35 cm²**
    (b) $\angle ACE = 180° − 50° = \mathbf{130°}$ (inside ∠'s //)
    (c) $\angle ECD = 180° − 130° = 50°$ (∠'s on st. line)
       $\angle CED = 90° − 50° = \mathbf{40°}$ (right Δ)

Workbook

Review 10, pp. 93-96

1. (a) 2 x 29

2. (a) 6

3. (a) $\dfrac{4}{9} \div 6 = \dfrac{\cancel{4}^{2}}{9} \times \dfrac{1}{\cancel{6}_{3}} = \dfrac{2}{27}$

   (b) $\dfrac{4}{\cancel{9}_{3}} \times \cancel{6}^{2} = \dfrac{8}{3} = 2\dfrac{2}{3}$

4. (a) 18.12   (b) 2163.8   (c) 58

5. (a) 1.13   (b) 0.56   (c) 90.63

6.

Decimals	0.5	0.8	**0.75**	**0.35**	0.48
Fractions	$\dfrac{1}{2}$	$\dfrac{4}{5}$	$\dfrac{3}{4}$	$\dfrac{7}{20}$	$\dfrac{12}{25}$
Percentage	50%	**80%**	**75%**	35%	**48%**

7. (a) $\dfrac{7}{20}$ x 100% = **35%**

   (b) 100% − 35% = **65%**

8. $2 = 200¢

   $\dfrac{80}{200}$ x 100% = **40%**

9. (a) 10% of $250 = **$25**

   (b) 75% of $1400 = $\dfrac{3}{4}$ x $1400 = **$1050**

10. Percentage left: 100% − 35% = 65%

    Amount left: 65% of $240 = $\dfrac{\cancel{65}^{13}}{\cancel{100}_{\cancel{20}_{1}}} \times \$\cancel{240}^{12}$ = **$156**

11. Increase: 15% of $160 = $\dfrac{\cancel{15}^{3}}{\cancel{100}_{\cancel{20}_{1}}} \times \$\cancel{160}^{8}$ = $24

    Selling price: $160 + $24 = **$184**

12. Percentage left: 100% − 25% − 20% = 55%

    Amount left: 55% of $12 = $\dfrac{\cancel{55}^{11}}{\cancel{100}_{\cancel{20}_{5}}} \times \$\cancel{12}^{3}$

    = $\dfrac{33}{5}$ = **$6.60**

    Or: 55% is 50% + 5%, or one half plus half a tenth. Half of $12 is $6. A tenth of $12 is $1.20 and half of that is 60¢. So 55% of $12 is $6 + $0.60 = $6.60.

13. (a) Cost of book: 20% of $50 = $\dfrac{1}{5}$ x $50 = **$10**

    (b) Remainder: $50 − $10 = $40

    Cost of magazine: 15% of $40 = $\dfrac{\cancel{15}^{3}}{\cancel{100}_{\cancel{20}_{1}}} \times \$\cancel{40}^{2}$

    = **$6**

    Or: 15% of $40 is one tenth plus half of a tenth of $40 = $4 + $2 = $6

14. (a) ∠ACB = 55°   (iso. Δ)
    ∠x = 180° − 55° = **125°**   (∠'s on st. line)
    (b) ∠DBC = 60°   (equ. Δ)
    ∠y = 180° − 60° − 25° = **95°** (∠'s on st. line)
    (c) ∠z + 90° = 38° + 90°   (ext. ∠ of Δ)
    ∠z = **38°**

15. ∠BCD = 108°   (opp. ∠'s of //)
    ∠DCE = 180° − 108°= **72°**   (∠'s on st. line)

16. ∠PSR = **55°**   (opp. ∠'s of //)
    ∠RTS = **90°**   (∠'s on st. line)
    ∠TRS = 90° − 55° = **35°**   (right Δ)

17. 3rd angle in triangle:
    180° − 60° − 40°= 80°   (∠ sum of Δ
    Angle opposite angle
    marked with x = 80°   (vert. opp. ∠'s)
    ∠x = 360° − 80° − 95° − 110°
    = **75°**   (∠ sum of ◊)

18. 1 kg of prawns: $12.75
    4 kg prawns: $12.75 x 4 = $51
    3 kg of fish: $76.50 − $51 = $25.50
    1 kg of fish: $25.50 ÷ 3 = **$8.50**

19. Number sold at $3 each: $\dfrac{4}{5}$ x 60 = 48
    48 x $3 = $144
    Number sold at 3 for $1: 60 − 48 = 12
    Groups of 3: 12 ÷ 3 = 4; 4 x $1 = $4
    Total money: $144 + $4 = **$148**

20. Lina starts with 1 unit and Suling with 2 units. For the ratio to become 2 : 1, Lina's units must increase to 4 (4 : 2 = 2 : 1)
    3 units = 24
    1 unit = 24 ÷ 3 = **8**
    Lina had 8 books at first.

    Lina
    Suling
    24

Unit 11 – Average and Rate

Chapter 1 – Average

Objectives

- Find the average of a set of data.
- Find the average when given the total and the number of items.
- Find the total when given the average and the number of items.
- Solve problems that involve averages and measurement in compound units.
- Solve word problems of up to 3 steps that involve averages.

Material

- Counters
- Multilink cubes

Vocabulary

- Average
- Per
- Set of data

Notes

In this chapter your student will learn to find the average of a set of data and solve problems involving averages that require not only finding the average, but also using the concept of averages to find data values.

The **average** is the arithmetical mean of a set of data.

A **set of data** is any list consisting of numerical values for individual or grouped items. Data can be information about individuals, such as the weight or height of each student in a classroom. Data can also be information about groups, such as the number of children in families, or the number of stores their families shop in, or the number of rainy days in different months.

To find the average for a set of data we first find the sum of the data values for each item on the list and then divide that sum by the number of items on the list.

For example, if 4 boys' ages are 10, 13, 11, and 9, their average age is found by first adding the four ages together and then dividing by 4. The boys' average age is 10.75 while their actual ages vary. Note that an average does not have to be a whole number, even though the list's items may all be whole numbers.

The expression for summing and dividing is sometimes written as a fraction.

Average = Total of items ÷ Number of items

Total: $10 + 13 + 11 + 9 = 43$
Average: $43 ÷ 4 = 10.75$ years
The average of 10 , 13, 11, and 9 is 10.75.

$$\frac{10 + 13 + 11 + 9}{4} = 10.75$$

If we are given the set of data's average and the number of items in the set of data we can find the total by multiplying the average by the number of items.

For example, if the average age of 4 boys is 10.75, we can find the sum of their ages by multiplying the average age by the number of boys.

> Total of items = Average x Number of items
>
> Average: 10.75
> Number of items: 4
> Total: 10.75 x 4 = 43

In *Primary Mathematics* 3 students learned to add and subtract in compound units and in *Primary Mathematics* 4 they learned to multiply or divide compound units. In this chapter your student will be solving problems involving measurement in compound units.

To multiply compound units we can multiply the different units separately, and then carry out any necessary conversions, as shown in the example at the right.

> It takes an average of 2 min 45 s to cycle 1 km. How long does it take to cycle 3 km?
> Total cycling time is 2 min 45 s x 3.
> Multiply minutes: 2 min x 3 = 6 min
> Multiply seconds: 45 s x 3 = 135 s
> Convert: 135 s = 2 min 15 s
> Add: 6 min + 2 min 15 s = 8 min 15 s
> The total cycling time is 8 min 15 s.

To divide compound units we can first divide the larger unit, then convert any remainder into the smaller unit, then add the smaller units together, and then divide these smaller units, as shown in the example at the right.

> It took 15 minutes 20 seconds to cycle 2 km. What is the average time per kilometer?
> Average time per km is 15 min 20 s ÷ 2.
> Divide minutes: 15 min ÷ 2 = 7 min R 60 s
> Add seconds together: 60 s + 20 s = 80 s
> Divide seconds: 80 s ÷ 2 = 40 s
> Add together: 7 min + 40 s = 7 min 40 s
> The average time per km is 7 min 40 s.

Although it is possible to convert everything to the smaller unit, multiply or divide, and then convert back to compound units, doing each part separately makes computation easier and allows for use of more mental math strategies, particularly with U.S. customary measurements.

As you do this unit have your student start recording the high and low temperatures for each day. At the end of this unit she can compute the average high and low temperatures for the period. Continue collecting the data, and save it to use in Unit 12.

(1) Find the average of a set of numbers

Activity

Draw 5 circles on some paper and put 2, 3, 4, 5, and 6 counters in them. Ask your student to find the total number of counters. Then ask him to rearrange the counters so that each circle has the same number of counters. Tell him that the number in each circle gives us the *average* number of counters. Point out that the total does not change.

Ask your student how she might find the average if she could not physically rearrange the objects. She can find the total and divide by the number of groups. The average is the number that would be in each circle if the total number, 20, were divided evenly among the 5 circles.

Draw a bar and divide it up into parts to show 2, 3, 4, 5, and 6. Then draw another bar below it and divide it up into 5 equal units. Tell your student that when we find the average, we are "evening" out the amount in each part. When each number is replaced by the average number, the sum of the numbers remains unchanged.

Use multilink cubes. Have your student link them together in a set of 4, a set of 6, and a set of 11, using a different color for each set. Ask him to find the average, both by rearranging the cubes and by finding the total and dividing.

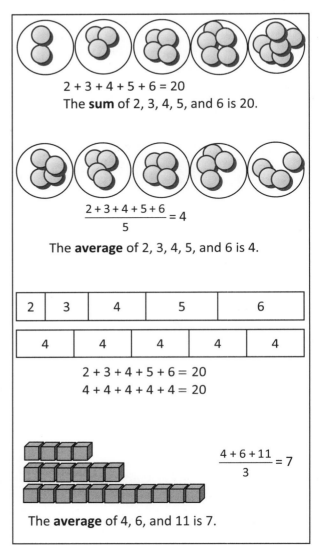

$$2 + 3 + 4 + 5 + 6 = 20$$
The **sum** of 2, 3, 4, 5, and 6 is 20.

$$\frac{2 + 3 + 4 + 5 + 6}{5} = 4$$

The **average** of 2, 3, 4, 5, and 6 is 4.

2	3	4	5	6

4	4	4	4	4

$$2 + 3 + 4 + 5 + 6 = 20$$
$$4 + 4 + 4 + 4 + 4 = 20$$

$$\frac{4 + 6 + 11}{3} = 7$$

The **average** of 4, 6, and 11 is 7.

Provide additional examples with linking cubes or counters. If your student has enough concrete experience, she will be able to find some simple averages mentally. For example, to find the average of 8, 10, and 12, she could mentally "see" that it would be 10 right away by taking 2 "off" of the 12 and putting it "on" the 8. Or, she can find the average of 4, 7, and 7 by mentally taking 1 off from each of the 7's and putting it onto the 4.

Discussion

Concept p. 106

Point out that in both cases, before the oranges are rearranged and after, the sum is the same. In both cases the average is also the same. The average is simply the total number divided by the number of groups; it says nothing by itself about the original numbers. An average of 6 could also mean there were 4, 9, and 5 in each bag, or 0, 9, and 9 in each bag.

Tasks 1-2, p. 107

Task 2: Have your student tell you how he could find the average without finding the total. He can just imagine 2 of the fish moved from Christian's row to Jamal's row. Then all boys would have the same number of fish, 5.

1. 5	5
2. 5	5

Activity

List some numbers and have your student find the average. Then discuss the following concepts:

⇒ The average is not higher than the highest number nor lower than the lowest number.

⇒ If more of the data values are at the higher end of the range of values, the average will be higher also.

⇒ An "outlying" value can draw the average down or up. For example, if the numbers of trees in 4 yards are 0, 8, 10, and 12, the average is 7.5. This is less than 3 out of the 4 values, because the far-out value 0 pulls the average down.

	Average
4, 5, 3, 5, 4	4.2
1, 2, 3, 2, 10	3.6
0, 8, 10, 12	7.5
1.2, 4.5, 6.7, 9.2	5.4
100, 82, 16	66

⇒ Just because the average number of trees is 7.5, that does not mean that there are half-trees. We may hear that the average number of children per family in some region is 2.3, but we know there isn't a single three tenths of a child running around anywhere. It just tells us that most families have two children, but a few have more.

Practice

Task 3-6, p. 108

Workbook

Exercise 1, pp. 97-100 (answers p. 116)

3. 36 + 38 + 40 = **114**
   Average number of stamps collected = **38**

4. (a) 1.4 m + 1.8 m + 2 m + 2.6 m + 3.2 m = **11 m**
   (b) Average length = $\dfrac{11 \text{ m}}{5}$ = **2.2 m**

5. (a) 68 + 76 + 78 + 88 = **310**
   (b) Average score = $\dfrac{310}{4}$ = **77.5**

6. Average distance = 1659 km ÷ 7 = **237 km**

(2) Solve problems involving averages

Activity

Draw 3 circles and tell your student that there are counters in each circle such that the average number of counters is 6, but you don't know exactly how many is in any one circle. There could be 9 in one, 5 in another, and so on. Ask her to see if she can find the total number of counters. She needs to multiply the average (6) by the number of groups (3). The total number of counters will be 18. Remind her that knowing the average and number of groups does not tell us anything about how many are actually in each group, but if they were distributed evenly into each group, there would be 6 in each, so we can use that value to find the total.

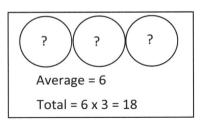

Average = 6

Total = 6 x 3 = 18

Practice

Tasks 7-8, p. 108

Task 6: Tell your student that the word *per*, as in per day, means "for each."

7. Total = 74.6 x 5 = **373**

8. Total = $4.65 x 8 = **$37.20**

Discussion

Tasks 9-10, p. 109

Use these two tasks to review multiplying and dividing compound units.

Task 9: Multiply each part by 3 separately. Convert the smaller unit to compound units and add the parts to get the final answer.

Task 10: Divide the larger unit, 5 kg, by 4 first. Convert the remainder of 1 kg to the smaller unit and add to the smaller unit. Then divide the sum by 4. Combine both parts for the final answer.

9. 4 kg 200 g

10. 1 kg 300 g

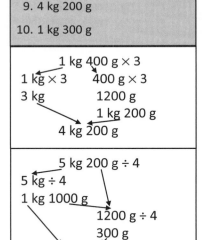

Practice

Tasks 11-12, p. 109

Workbook

Exercise 2, pp. 101-102 (answers p. 116)

11. 7 min 40 s

12. 8 min 15 s

(3) Solve problems involving averages

Discussion

Tasks 13-14, p. 109

You can draw circles to help students visualize the problems, as shown below. Each circle represents different data values.

13. 1.48 m
    1.48 m

14. $5.70
    $5.70

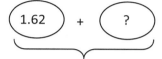

Average height of 2 boys: 1.55 m
Total height: 1.55 m × 2 = 3.10 m
Height of other boy:
3.10 m − 1.62 m = 1.48 m

Average cost of 2 books: $3.90
Total cost of 2 books: $3.90 × 2 = $7.80

Average cost of 3 books: $4.50
Total cost of 3 books: $4.50 × 3 = $13.50

Cost of third book: total cost − cost of 2 books
$13.50 − $7.80 = $5.70

Practice

Practice A, p. 110

Workbook

Exercise 3, p. 103 (answers p. 116)

Reinforcement

Extra Practice, Unit 11, Exercise 1, pp. 235-238

Test

Tests, Unit 11, 1A and 1B, pp. 139-142

1. (a) 26.3
   (b) $2.81
   (c) 4.1 kg
   (d) 5.15 ℓ
   (e) 2.39 m
   (f) 19.4 km
   (g) 3.48 gal
   (h) 7.77 in.

2. 5460 km ÷ 3 = **1820 km**
   He traveled an average of 1820 km per month.

3. 18 kg x 6 = **108 kg**
   The total weight of the 6 packages is 108 kg.

4. $3.75 x 4 = **$15**
   The total cost of the lunch was $15.

5. 1 h 20 min x 5 = (1 x 5) h + (20 x 5) min = **6 h 40 min**
   She spent a total of 6 h 40 min reading storybooks.

6. 10 ℓ 275 ml ÷ 3= **3 ℓ 425 ml**
   He used an average of 3 ℓ 425 ml of gas per day.

7. Total cost of 2 books: $2.45 x 2 = $4.90
   Cost of other book: cost of 2 books − cost of 1 book
   $4.90 − $2.80 = **$2.10**
   The other book cost $2.10.

8. Number of people for first three days: 145 x 3 = 435
   Number of people for four days: 435 + 205 = 640
   Average per day: 640 ÷ 4 = **160**
   There is an average of 160 visitors per day.

Workbook

Exercise 1, pp. 97-100

1. (a) The sum is **18**.
   18 ÷ 3 = **6**
   The average is **6**.
   (b) (45 + 33) ÷ 2 = 78 ÷ 2 = **39**
   (c) (24 + 38 + 19) ÷ 3 = 81 ÷ 3 = **27**
   (d) (20 + 18 + 36 + 98) ÷ 4 = 172 ÷ 4 = **43**

2. (9 + 6 + 5 + 8) ÷ 4 = 28 ÷ 4 = **7**
   Each boy made an average of 7 kites.

3. ($25 + $18 + $32 + $29) ÷ 4 = $104 ÷ 4 = **$26**
   Their average savings is $26.

4. (a) ($3.70 + $4.25 + $4.50) ÷ 3
       = $12.45 ÷ 3
       = **$4.15**
   (b) (12.5 m + 14.7 m + 12.4 m) ÷ 3
       = 39.6 m ÷ 3
       = **13.2 m**
   (c) (15.5 kg + 12 kg + 14.3 kg + 16.6 kg) ÷ 4
       = 58.4 kg ÷ 4
       = **14.6 kg**
   (d) (430 ℓ + 22 ℓ) ÷ 2
       = 452 ℓ ÷ 2
       = **226 ℓ**

5. (3.8 yd + 5 yd + 5.42 yd + 4.5 yd) ÷ 4
   = 18.72 yd ÷ 4
   = **4.68 yd**
   The average distance is 4.68 yd.

6. (1.2 lb + 0.85 lb + 1.5 lb + 1.75 lb + 1.4 lb) ÷ 5
   = 6.7 lb ÷ 5
   = **1.34 lb**
   The average weight is 4.68 lb.

Exercise 2, pp. 101-102

1. 258 ÷ 3 = **86**
   He sold an average of 86 plums per day.

2. 720 g ÷ 8 = **90 g**
   The average weight is 90 g.

3. 12.4 x 3 = **37.2**
   The sum of the numbers is 37.2.

4. 28.5 in. x 4 = **114 in.**
   The total length is 114 in.

5. (a) 2 ℓ 450 ml x 2 = (2 ℓ x 2) + (450 ml x 2)
       = **4 ℓ 900** ml
   (b) 2 m 65 cm x 3 = (2 m x 3)+ (65 cm x 3)
       = 6 m + 195 cm = **7 m 95 cm**
   (c) 6 km 250 m x 5 = (6 km x 5) + (250 m x 5)
       = 30 km + 1250 m = **31 km 250 m**
   (d) 3 kg 300 g ÷ 3 = (3 kg ÷ 3) + (300 g ÷ 3)
       = **1 kg 100 g**
   (e) 5 h 30 min ÷ 3
          5 h ÷ 3 = 1 h R 2 h = 1 h 120 min
          120 min + 30 min = 150 min
          150 min ÷ 3 = 50 min
       = **1 h 50 min**
   (f) 1 ℓ 600 ml ÷ 4 = 1600 ml ÷ 4 = **400 ml**
   (g) 4 lb 3 oz x 6 = (4 lb x 6) + (3 oz x 6)
       = 24 lb + 18 oz = 24 lb + 1 lb 2 oz = **25 lb 2 oz**
   (h) 2 ft 10 in. x 4 = (2 ft x 4) + (10 in. x 4)
          = 8 ft + 40 in. = 8 ft + 3 ft 4 in. = **11 ft 4 in.**

6. 9 kg 400 g ÷ 4
      9 kg ÷ 4 = 2 kg R 1 kg = 2 kg R 1000 g
      1000 g + 400 g = 1400 g
      1400 g ÷ 4 = 350 g
   = **2 kg 350 g**
   The average weight is 2 kg 350 g.

7. 2 ℓ 250 ml x 6 = (2 ℓ x 6) + (250 ml x 6)
   = 12 ℓ + 1 ℓ 500 ml = **13 ℓ 500 ml**
   The total amount of water is 13 ℓ 500 ml.

Exercise 3, p. 103

1. Total weight of Ali, Mick, and Samy:
   45 kg x 3 = 135 kg
   Weight of Ali and Mick: 85 kg
   Weight of Samy: 135 kg − 85 kg = **50 kg**
   Samy weighs 50 kg.

2. Total spent Mon. to Sat.: $4.50 x 6 = $27.00
   Total spent Mon. to Sun.:
   $27.00 + $5.20 = $32.20
   Average spent per day Mon. to Sun.:
   $32.20 ÷ 7 = **$4.60**
   The average amount of money spent from
   Monday to Sunday was $4.60.

Chapter 2 – Rate

Objectives

♦ Understand rate as one quantity per unit of another quantity.
♦ Solve word problems that involve rate.

Material

♦ Appendix pp. a14-a17

Vocabulary

♦ Rate
♦ Per

Notes

In this chapter your student will be introduced to rate.

A **rate** involves two quantities that correspond to each other. It is usually expressed as one quantity (or measurement) per unit of another quantity (or measurement).

Given two quantities A and B such that 5 units of A corresponds to 1 unit of B, we say that the rate is 5 units of A per unit of B. The word **per** means "for every" and can also be written with a /.

Speed is an example of rate. If a car goes 100 kilometers in 1 hour, the rate is 100 kilometers per hour, 100 km/h. The cost of something by weight is another example of rate. If meat costs $6 for each kilogram, we can say that the rate is $6 per kilogram. How fast a person types is another example of rate.

For whole number, fraction, and ratio problems students learned to set up bar models to diagram the pertinent information. Often, the solution involves initially finding the value of 1 unit. This unitary approach is used in solving many word problems including percentage problems. This underlying approach will also be used for solving rate problems. Rather than drawing bar models though, your student will learn how to use a line model to diagram the information in word problems that involve rate. The line model will help your student understand and set up the problem correctly, and, after doing the computations, see at a glance whether their answer is reasonable. The activities in these lessons relate line models to the familiar bar models, and finding the rate for a unit quantity to finding the value of one unit in a bar model.

⇒ Simon can type 200 words in 5 minutes. How many words can he type in 3 minutes?

A number line with both types of quantities might look like the figure at the right.

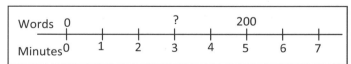

The markings are labeled for minutes on one side of the line and for words on the other side. Since 200 words are typed in 5 minutes, the mark for 200 words is also the one for 5 minutes. The question mark for the number of words typed in 3 minutes is at the same mark as the 3-minute mark. When we know the scale for the top part of the number line, we can find the value at the point indicated by the question mark. In this problem we can find the scale by dividing 200 by 5, which gives us the number of words that can be typed in 1 minute (40). From there we can find the number of words typed in 3 minutes by multiplying that by 3 (40 x 3 = 120).

In a line model only two marks are used, one for the two known quantities linked in the problem (here, 200 words and 5 minutes) and one for the quantity that needs to be found (indicated with "?") and its linked quantity (here, 3 min). The smaller quantities (3 min. and ? words) are put at the left mark, and the larger quantities (5 min and 200 words) at the right mark.

Using a line model will help your student to write an arrow diagram. The arrow diagram shows the relationship between the two quantities while the line diagram helps him determine which quantity to put on the left side of the arrow diagram. The arrow symbolizes the words "corresponds to" or "per." With the arrow diagram, your student can determine whether to multiply or divide and by how much.

To draw an arrow diagram for this example:

Put minutes on the left side of the arrow, since we want to find the number of words he types in 1 minute. This is the "unitary" approach to solving rate problems.

In one minute he would type a fifth as many words, so divide 200 by 5. Notice that we carrying out the same operation on the values on both sides of the arrow — we divide the number of minutes by 5 and the number of words by 5.

Then find the value for the number of words typed in 3 minutes. He would type 3 times as many words in 3 minutes as in 1 minute, so we multiply by 3. Again, we are carrying out the same operation on the values on both sides of the arrow.

When your student understands the process, the steps in the arrow diagram can be combined as shown at the right.

> Simon can type 200 words in 5 minutes. How many words can he type in 3 minutes?
>
> ? words   200 words
> |————————————|
> 3 min   5 min

> 5 min $\longrightarrow$ 200 words
>
> 1 min $\longrightarrow$ $\dfrac{200}{5}$ words
>
> 3 min $\longrightarrow$ $\dfrac{200}{5}$ x 3 words
>
>   = 120 words
>
> He can type 120 words in 3 minutes.
>
> 5 min $\longrightarrow$ 200 words
>
> 3 min $\longrightarrow$ $\dfrac{200}{5}$ x 3 words

Your student will also have problems involving rate tables. With rate tables the rate increases in steps, rather than continuously. For example, if parking costs $1 for each hour, the parking fee for 1 h 20 min is the same as the fee for 2 hours.

(1) Understand rates

Discussion

Concept p. 111

Before looking at this page write the following easy problem and discuss its solution with an appropriate bar model, as shown at the right.

⇒ Mrs. Sweet bought 60 bottles of syrup and packed them into 5 boxes. How many bottles were put in each box?

Then read the problem on p. 111 of the textbook.

⇒ A machine fills 60 similar bottles of syrup in 5 minutes. How many such bottles of syrup can it fill in one minute?

Point out that this problem is similar to the previous one, but instead of finding the number of bottles for each box, we are finding the number the bottles filled for each minute. We could use a similar bar diagram with 1 unit bar for the bottles filled in 1 minute. We need 5 units for 5 minutes. To find the number of bottles filled each minute, we divide the number of bottles filled (60) by the number of minutes it takes (5). Tell your student that the answer is the *rate* for the number of bottles filled each minute. *Rate* is the relationship between two quantities with different measurement units. In this problem, the two units are bottles and minutes. We use the word *per* for rate to mean "for each." We say that the rate at which the machine fills the bottles is 12 bottles per minute. Sometimes "per" is written as "/". We read 12 bottles/minute as "12 bottles per minute."

Ask your student to find the number of bottles filled in 3 minutes. Once we know the rate of 12 bottles/min, all we need to do is multiply by 3 to find the number of bottles filled in 3 minutes.

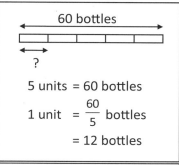

60 bottles

?

5 units = 60 bottles

1 unit $= \dfrac{60}{5}$ bottles

$= 12$ bottles

60 bottles

1 min

? bottles

Rate $= \dfrac{60}{5} \dfrac{\text{bottles}}{\text{min}}$

$= \dfrac{12}{1} \dfrac{\text{bottles}}{\text{min}}$

$= 12$ bottles per minute

$= 12$ bottles/minute

How many bottles are filled in 3 minutes?

12 bottles/min x 3 min
= 36 bottles

Discuss other instances where rate is used. Often rate is used for speed, such as how fast a person can type in words per minute or how fast a car goes in kilometers per hour or miles per hour. Rate is also used for interest and tax rate, as seen in the unit on percentage. For example, the interest rate could be 5% per year. In interest and tax rates, one quantity is the percentage of the money invested or earned, and the other is time, often 1 year. Rate is not just used with time, for example, it is also used in how many miles a car gets per gallon.

Practice

Task 1-4, p. 112

Workbook

Exercise 4, pp. 104-105 (answers p. 125)

1. $5
2. 25   25
   25
3. 720   720
   720
4. 150   150
   150

(2) Solve rate problems

Activity

Write the following problem:

⇒ Anna makes $15 an hour. How much does she make in 6 hours?

Tell your student that this is a simple problem, but we are just going to use it to show how line models and arrow diagrams help with solving rate problems so they can be used with harder problems.

Draw a bar model as shown at the right, Below it draw a line to start the line model. Explain what you are doing as you draw the model.

First make a mark on the line and label it for the rate relationship that we know: $15 and 1 hour. Usually time is put on the bottom. Then make another mark for the relationship that has an unknown value, the amount of money for 6 hours. Since 6 hours is greater than 1 hour, we mark 6 hours to the right of the first mark. Label the mark with "?" for the unknown number of dollars.

Then write the relationship between the two quantities as shown on the right. The number of dollars she makes in 1 hour is $15. Tell your student that rather than write out "number of dollars" we can just write 1 h, but 1 h does not really equal $15, so instead of an equal sign we use an arrow. The arrow simply connects the two quantities. In this case the arrow would mean "...of pay is..."

To go from 1 hour to 6 hours we multiply 1 by 6. To go from $15 to the amount of pay in 6 hours, we also multiply by 6. So in place of the question mark in the arrow diagram, we can write the equation we will use to find the answer and then find the answer.

So why bother with line model rather than just drawing a bar diagram? With a line model all we need to make is 2 marks at either end for the four values in a rate problem, the three we know and the one we need to find. If we had been asked for the amount of money Anna makes in 30 hours instead, and wanted to draw a diagram to show the relationship between the four values, we don't need to show the 30 hours as individual units, which would be tedious to draw. This particular problem is easy, but a quick line model can help with more difficult problems.

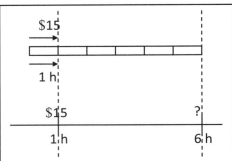

Number of dollars in 1 hour = $15
Number of dollars in 6 hours = ?

$$1\,h \longrightarrow \$15$$
$$6\,h \longrightarrow ?$$

$$\times 6 \left(\begin{array}{l} 1\,h \longrightarrow \$15 \\ 6\,h \longrightarrow \$15 \times 6 \\ = \$90 \end{array} \right) \times 6$$

Use appendix p. a14 and discuss the following problems. Have your student draw a line model and use an arrow diagram for each.

1. A machine makes 1800 cans of soda pop in 1 day (24 hours). How many cans does it make in 1 hour?

To go from 24 hours to 1 hour we divide by 24, so we also divide 1800 cans by 24. This lets us find the number of cans made in 1 hour.

Tell your student that this is a good time to get into the practice of writing division as fractions. In many cases he will be able to simplify the fraction before dividing. In this case if we simplify the fraction in steps, we avoid dividing 1800 directly by 24.

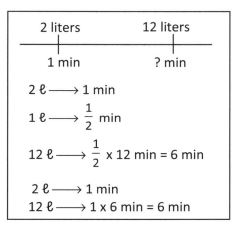

2. Water flows from a tap at the rate of 2 liters per minute. How long will it take to fill a 12 liter tank?

We want to find the new time for a given number of liters, so we put liters first in the arrow diagram. We can solve first for the number of minutes for 1 liter and then for 12 liters. Or, since 12 liters is 6 times as much as 2 liters, we could simply multiply the number of minutes by 6.

$\frac{1}{2}$ min/ℓ is a rate, the number of minutes per liter, even though rates are more often given per unit of time rather than the other way around.

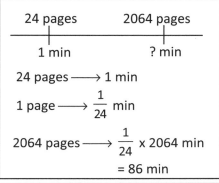

3. A photocopier can print 24 pages per minute. How long does it take to print 2064 pages?

We know the two values for the number of pages, so we put pages to the left of the arrows, and find the number of minutes for 1 page.

Practice

Tasks 5-6, p. 113

Task 5(b): Point out that we could solve this in one step. Since 100 gal is 4 times 25 gal, we could multiply 1 min by 4.

Task 6: We could also solve this in one step, but it is less obvious that 135 is 3 times 45, so in this case it is easier to show that we divide by 45 and multiply by 135 and then simplify $\frac{135}{45}$.

Workbook

Exercise 5, pp. 106-107 (answers p. 125)

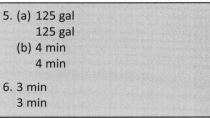

5. (a) 125 gal
      125 gal
   (b) 4 min
      4 min

6. 3 min
   3 min

(3) Solve rate problems

Discussion

Tasks 7-8, p. 114

You may want to write the problems on paper or whiteboard and discuss the solution, having your student attempt a line model and arrow diagram before looking at the solution in the textbook. These problems do show the usefulness of starting with a line drawing to help with determining how to set up an arrow diagram.

7. (a) 12 km
180 km
180 km
(b) 10 ℓ
10 ℓ
8. 15
15

Task 7(a): We know both numbers for liters and want to find the new value for kilometers, so liters goes on the left side for the arrow diagram.

Task 7(b): This time we know both values for kilometers, so that is what goes on the left side for the arrow diagram. You might want to point out that if we solved the intermediate step to find the rate of liters per kilometers, we would get a non-terminating decimal. So it is better to leave the calculations to the end, where we can simplify.

Practice

Appendix p. a15

1. A machine can make 480 toys in an hour. How many toys can it make in 15 minutes?

   60 min ⟶ 480 toys
   15 min ⟶ 480 toys ÷ 4 = 120 toys

2. It takes 10 hours for a carpenter to make 5 chairs. How long will it take him to make 182 chairs?

   5 chairs ⟶ 10 h
   182 chairs ⟶ $\frac{10}{5}$ h x 182 = 364 h

3. A machine can produce 65 cans of soda in 4 minutes. How long will it take to produce 1300 cans?

   65 cans ⟶ 4 min
   1300 cans ⟶ $\frac{4}{65}$ min x 1300 = 80 min

4. A wheel makes one third of a revolution in one second. How long does it take to make 1000 revolutions?

   $\frac{1}{3}$ rev ⟶ 1 s
   1 rev ⟶ 3 s
   1800 rev ⟶ 3 s x 1800 = 5400 s = 90 min

5. Sara saves 35¢ a day. How many days will she take to save at least $10?

   35¢ ⟶ 1 day
   1000¢ ⟶ $\frac{1}{35}$ day x 1000 = $28\frac{4}{7}$ days
   She needs to save for at least 29 days.

6. A photocopier can print 16 pages a minute. How long will it take to print 9 copies of a document that is 256 pages long?

   16 pages ⟶ 1 min
   9 x 256 pages ⟶ $\frac{1}{16}$ min x 9 x 256
   = 144 min

Workbook

Exercise 6, pp. 108-109 (answers p. 125)

(4) Solve rate problems involving steps

Discussion

Task 9, p. 115

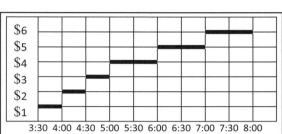

9. $9
   $9

After discussing the problem in the text, ask your student what the total parking fee would be if the car had been parked there from 3:30 pm to 7:30 pm.

The fee from 3:30 to 5:00 is $1 for each half hour. There are 3 half hours, so the fee for that portion of the time is $3. From 5:00 pm to 7:30 pm is two and a half hours. The fee is $1 per hour during that time. So the fee for that time would be $3, because any portion of an hour after 5:00 pm costs $1. The fee is $1 from 7:00 to 8:00, no matter what time the car leaves between 7:00 and 8:00. The rate goes up in steps for each half hour or hour, not continuously.

Have your student find the parking fees for other durations at different times of the day.

Write the following problem and discuss its solution.

⇒ The rate for telephone calls to another country is $12.50 for the first block of 3 minutes, and $1.80 for each subsequent block of 30 seconds.

How much does Jasmine have to pay for an $8\frac{1}{4}$ minute call to this country?

> The first 3 minutes cost $12.50.
> There are 10 half-minutes from 3 min to 8 min, and the last quarter minute counts as a half-minute.
> $1.80 x 11 = $19.80
> $12.50 + $19.80 = $32.30
> She has to pay $32.30.

If she has $50.00 for a telephone call to her aunt in this country, what was the longest time in minutes Jasmine could talk to her aunt?

> $50.00 − $12.50 = $37.50
> $37.50 ÷ $1.80 = ($36 + $1.50) ÷ $1.80
> = 20 R $1.50
> 3 min + 20 half-minutes = 13 min
> She could talk for at most 13 minutes.

Practice

Tasks 10-12, pp. 115-116

Task 11: Your student should understand that the postage rates are in steps. Any magazine between 51 g and 100 g costs $0.70 to mail. The rate for each step is shown in the right column for rates up to 100 g. But for weights greater than 100, the cost is $0.70 for the first 100 g, and then an *additional* $0.60 on top of the $0.70 for every additional 100 g or portion thereof.

Workbook

Exercise 7, pp. 110-111 (answers p. 125)

10. $84
    $76
    $84 + $76 = **$160**
    $160

11. (a) $0.70
        $0.70
    (b) $1.80
        $0.70 + $1.80 = **$2.50**
        $2.50

12. $2
    $2.40 + $2 = **$4.40**
    $4.40

(5) Practice

Activity

Your student may already have experience with measuring rates in sports or science experiments. If not, you may want to let him measure his typing rate in words per minute. Have him type a given passage for a set amount of time using a word processing program that determines word count, or time how long it takes to type the passage. Use the word count feature of the program to find the number of characters including spaces in the passage, and divide it by 5 to find the number of words (most typing programs use 5 characters per word as an average). Then have him find the rate of words per minute.

Practice

Practice B, p. 117

Reinforcement

Extra Practice, Unit 11, Exercise 8, pp. 239-242

Test

Tests, Unit 11, 2A and 2B, pp. 143-146

Enrichment

Appendix p. a16

(answers on appendix p. a17)

---

1. 50 pages $\longrightarrow$ 1 min

   2500 pages $\longrightarrow$ $\frac{1}{50}$ min x 2500 min = **50 min**

   It will take 50 min to print 2500 pages.

2. 4 min $\longrightarrow$ 16 boxes

   1 min $\longrightarrow$ $\frac{16}{4}$ boxes = **4 boxes**

   It can seal 4 boxes in 1 minute.

3. 2 min $\longrightarrow$ 152 beats

   30 min $\longrightarrow$ $\frac{152}{2}$ beats x 30 = **2280 beats**

   It beats 2280 times in 30 minutes.

4. 100 gal $\longrightarrow$ 5 min

   1000 gal $\longrightarrow$ $\frac{5}{100}$ min x 1000 min = **50 min**

   It will take 50 min to fill the pool.

5. 30 m^2 $\longrightarrow$ $810

   55 m^2 $\longrightarrow$ $\$\frac{810}{30}$ x 55 = **$1485**

   It will cost $1485.

6. 25 rev $\longrightarrow$ 40 m

   50 rev $\longrightarrow$ $\frac{40}{25}$ m x 50 = **80 m**

   It will cover 80 m in distance.

7. (a) 55 g is greater than 40 g but less than 100 g. Postage is **$0.50**.
   (b) 400 g is greater than 250 g but less than 500 g. Postage is **$1.00**.

8. (a) Rent for Fri.: $60
   Rent for Sat. and Sun.: $80 x 2 = $160
   Total rent: $60 + $160 = **$220**
   (b) Rent for Wed. to Fri.: $60 x 3 x 2 = $360
   Rent for Sat.: $80 x 2 = $160
   Total rent: $360 + $160 = **$520**

Workbook

Exercise 4, pp. 104-105

1. (a) **$75** per day   Rate $= \dfrac{225}{3} = 75

   (b) **50** words per minute   Rate $= \dfrac{750}{15} = 50$

   (c) **12** jars per minute   Rate $= \dfrac{240}{20} = 12$

   (d) **34** mi per gal   Rate $= \dfrac{102}{3} = 34$

2. (a) $44 \times 5 = $ **220**
   (b) $225 \times 35 = $**7875**
   (c) $25 \times 7 = $ **175**
   (d) $24 \text{ m}^3 \times 6 = $ **144 m^3**

Exercise 5, pp. 106-107

1. 100 words $\longrightarrow$ 1 min

   2000 words $\longrightarrow \dfrac{1}{100}$ min x 2000 = **20 min**

2. $80 $\longrightarrow$ 1 day

   $400 $\longrightarrow \dfrac{1}{80}$ day x 400 = **5 days**

3. 6 rev $\longrightarrow$ 1 min

   45 rev $\longrightarrow \dfrac{1}{6}$ min x 45 = **7.5 min**

4. 1 h $\longrightarrow$ $6
   7 h $\longrightarrow$ $6 x 7 = **$42**

5. 1 min $\longrightarrow$ 200 loaves
   5 min $\longrightarrow$ 200 loaves x 5 = **1000** loaves

6. 12 km $\longrightarrow$ 1 ℓ

   180 km $\longrightarrow \dfrac{1}{12}$ ℓ x 180 = **15 ℓ**

Exercise 6, pp. 108-109

1. $\dfrac{300}{5} = 60

   (a) 6 days $\longrightarrow$ $60 x 6 = **$360**
   (b) $300 $\longrightarrow$ 5 days

        $1200 $\longrightarrow \dfrac{5}{300}$ days x 1200 = **20 days**

2. $\dfrac{84}{6}$ km = **14** km

   (a) 16 ℓ $\longrightarrow$ 14 km x 16 = **224 km**
   (b) 84 km $\longrightarrow$ 6 ℓ

        210 km $\longrightarrow \dfrac{6}{84}$ ℓ x 210 = **15 ℓ**

3. $\dfrac{1600}{40} = 40

   (a) 90 m$^2 \longrightarrow$ $40 x 90 = **$3600**
   (b) $1600 \longrightarrow$ 40 m^2

        $2000 \longrightarrow \dfrac{40}{1600}$ m^2 x 2000 = **50** m^2

4. $\dfrac{600}{4} = $ **150**

   (a) 15 min $\longrightarrow$ 150 x 15 pages = **2250** pages
   (b) 600 pages $\longrightarrow$ 4 min

        750 pages $\longrightarrow \dfrac{4}{600}$ min x 750 = **5 min**

5. (a) 2 days $\longrightarrow$ 80 s

        3 days $\longrightarrow \dfrac{80}{2}$ s x 3 = **120 s**

   (b) 80 s $\longrightarrow$ 2 days

        200 s $\longrightarrow \dfrac{2}{80}$ days x 200 = **5 days**

6. (a) 12 min $\longrightarrow$ 1500 books

        5 min $\longrightarrow \dfrac{1500}{12}$ books x 5 = **625 books**

   (b) 1500 books $\longrightarrow$ 12 min

        1000 books $\longrightarrow \dfrac{12}{1500}$ min x 1000 = **8 min**

Exercise 7, pp. 110-111

1. (a) Fare for 1st hour = $3
   Fare for 2nd hour = $2
   Total fare = $3 + $2 = **$5**
   (b) She rented the bicycle for 4 hours.
   Fare of 1st hour = $3
   Fare for next 3 hours = $2 x 3 = $6
   Total fare = $3 + $6 = **$9**
   (c) Fare for 1st hour for 2 bicycles = $3 x 2 = $6
   Fare for next 2 hours = $2 x 2 x 2 = $8
   Total fare = $6 + $8 = $14
   $14 ÷ 4 = **$3.50**

2. (a) 15 m^3 is within the first 20 m^3.
   $0.56 per m^3 x 15 m^3 = **$8.40**
   (b) Charge for 1st 20 m^3 = $0.56 x 20 = $11.20
   Charge for next 10 m^3 = $0.80 x 10 = $8.00
   Total charge = $11.20 + $8.00 = **$19.20**
   (c) Charge for 1st 20 m^3 = $11.20
   Charge for next 20 m^3 = $0.80 x 20 = $16.00
   Charge for last 5 m^3 = $1.17 x 5 = $5.85
   Total charge = $11.20 + $16.00 + $5.85 = **$33.05**

Review 11

Review

Review 11, pp. 118-121

2: You may want to point out that measurement conversions are actually rate problems, and arrow diagrams are easy to use for them. Simply use an arrow to show the relationship between the two measurements, with the unit we are given on the left. It is then easy to see whether we need to multiply or divide. If 1 hour is 60 minutes, then $\frac{2}{3}$ h is 60 x $\frac{2}{3}$ min.

Another example: Find the number of hours for 25 minutes. In an arrow diagram put minutes first, since we need to find the number of hours for 25 minutes. It is easy to see that we need to divide.

60 min $\longrightarrow$ 1 h

1 min $\longrightarrow$ $\frac{1}{60}$ h

25 min $\longrightarrow$ $\frac{1}{60}$ h x 25 = $\frac{5}{12}$ h

4: The price of something per weight is also a ratio, and an arrow diagram makes it easy to see if we need to multiply or divide. Since we are finding the cost for $\frac{3}{4}$ kg, we put kg first in the arrow diagram for the relationship we are given, 1 kg costs $12.

Workbook

Review 11, pp. 112-115 (answers p. 128)

Tests

Tests, Units 1-11, Cumulative Tests A and B, pp. 147-154

---

1. (a) $2\frac{1}{4}$ kg = (2 x 1000) + ($\frac{1}{4}$ x 1000) g = **2250 g**

   (b) 3 km 90 m = 3 km + $\frac{90}{1000}$ km = **3.09 km**

   (c) $2\frac{3}{10}$ m = 2 m + ($\frac{3}{10}$ x 100) cm = **2 m 30 cm**

2. 1 h $\longrightarrow$ 60 min

   $\frac{2}{3}$ h $\longrightarrow$ 60 min x $\frac{2}{3}$ = 40 min

   The concert lasted 1 h 40 min

   7 h 15 min + 1 h 40 min = 8 h 55 min

   The concert ended at **8:55 p.m.**

3. $\frac{1}{4}$ of $\frac{2}{5}$ of the people are boys.

   $\frac{1}{4}$ x $\frac{2}{5}$ = **$\frac{1}{10}$**

   $\frac{1}{10}$ of the people are boys.

4. 1 kg $\longrightarrow$ $12

   $\frac{3}{4}$ kg $\longrightarrow$ $12 x $\frac{3}{4}$ = **$9**

   $\frac{3}{4}$ kg of beef costs $9.

5. $\frac{5}{8}$ of money $\longrightarrow$ $240

   $\frac{1}{8}$ of money $\longrightarrow$ $\frac{240}{5}$

   All of money $\longrightarrow$ $\frac{240}{5}$ x 8 = **$384**

   The sum of money is $384.

   Or

   5 units = $240

   8 units = $\frac{240}{5}$ x 8 = $384

   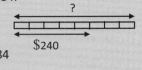

6. 12

7. 2×5^2

8. (a) $\frac{185+103+127+165}{4} = \frac{580}{4}$ = **145**

   (b) $\frac{3.8+2.7+4.5+1.6}{4}$ cm = $\frac{12.6}{4}$ cm = **3.15 cm**

9. Total = 40 x 5 = 200

   200 − (18 + 27 + 37 + 50) = 200 − 132 = **68**

   The fifth number is 68.

(continued next page)

10. (a) 1 : 5   (b) 1 : 2   (c) 3 : 2
    (d) 4 : 1   (e) 5 : 2   (f) 1 : 25

11. (a) 15   (b) 8

12.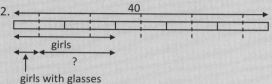

10 units = 40
1 unit = 40 ÷ 10 = 4
3 units = 4 x 3 = **12**   Or: $\frac{3}{4}$ x ($\frac{2}{5}$ x 40) = 12
12 girls do not wear glasses.

13.

14 x 36

□ = 1 unit   sold   ?

There are 6 larger units. Divide each into 6.
Total number of units = 6 x 6 = 36
5 units were left.
36 units = 14 x 36
1 unit = $\frac{14 \times 36}{36}$ = 14
5 units = 14 x 5 = **70**
He had **70** apples left.   Or: $\frac{5}{6}$ x ($\frac{1}{6}$ x 14 x 36) = 70

14. 3 units = $45
    handbag   belt and amount left
5 units = $ $\frac{45}{3}$ x 5 = **$75**
                  $20 + $25 = $45
She had **$75** at first.

15. 1 unit = 12   boys   girls
2 units = 12 x 2 = 24
There are **24** girls.   12

16. 4 units = 20   20
1 unit = 20 ÷ 4 = 5   red
(a) 3 units = 3 x 5 = **15**   green
    There are 15 green buttons.
(b) 7 units = 7 x 5 = **35**
    There are a total of 35 buttons.

17. Total units: 3 + 2 + 7 = 12; Total money: $156
12 units: $156
1 unit = 156 ÷ 12 = 13   Lily
(a) 2 units = $13 x 2 = **$26** Carla   ?
    Carla received $26.   Gwen
(b) Gwen received 4 more units than Lily.
    4 units = $13 x 4 = **$52**
    Gwen received **$52** more than Lily.

18. 96% of 500 = $\frac{96}{100}$ x 500 = **480**
She sold 480 cookies.

19. Percentage left = 100% - 35% = 65%
65% of $350 = $\frac{65}{100}$ x $350 = **$227.50**
She had $227.50 left.

20. Discount: 15% of $3600 = $\frac{15}{100}$ x $3600 = $540
Amount paid: $3,600 − $540 = **$3060**

21. Increase: 10% of $640 = $64
Selling price: $640 + $64 = **$704**

22. Side: 36 ÷ 4 = 9 cm
Area: 9 cm x 9 cm = **81 cm^2**

23. No problem 23 in 2008 printing.

24. Total length: 28 cm + (2 x 5 cm) = 38 cm
Total width: 25 cm + (2 x 5 cm) = 35 cm
Total area: 38 cm x 35 cm = 1330 cm^2
Area of photograph: 28 cm x 25 cm = 700 cm^2
Area not covered: 1330 cm^2 − 700 cm^2 = **630 cm^2**

25. ∠DCE = 90° − 58° = 32°   (right Δ)
∠ACB = ∠DCE = 32°   (vert. opp. ∠'s)
∠x = ∠ACB = **32°**   (iso. Δ)

26. ∠DCE = 360° − 95° − 62° − 83° = **120°**

27. Base: 53.3 cm^2 ÷ 6.5 cm = **8.2 cm**

28. Surface area:
2 x (14 in. x 4.5 in. + 14 in. x 6.5 in. + 4.5 in. x 6.5 in.)
= 2 x (63 in.^{2} + 91 in.^{2} + 29.25 in.^{2})
= 2 x 183.25 in.^{2} = **366.5 in.^{2}**

29. Volume = 400 cm^2 x 30 cm = **12,000 cm^3**

30. Total weight of 3 packages: 2.2 kg x 3 = 6.6 kg
Total weight of 2 packages: 1.8 kg x 2 = 3.6 kg
Weight of third packages: 6.6 kg − 3.6 kg = **3 kg**

31. (a) 50 cm x 30 cm x 20 cm = 30,000 cm^3 = **30 ℓ**
(b) 12 ℓ ⟶ 1 min
30 ℓ ⟶ $\frac{1}{12}$ x 30 min = **2$\frac{1}{2}$ min**

32. 20 ℓ ⟶ 1 min
800 ℓ ⟶ 1 x 40 min = **40 min**
It will take 40 min to fill the pool.

33. 800 ÷ 250 = 3 R 50
First 250 g: $20.00
Next 250 g + 250 g + 50 g: $2.80 x 3 = $8.40
Total cost: $20 + $8.40 = **$28.40**

Workbook

Review 11, pp. 112-115

1. (a) $120 - \underline{20 \div 5} = 120 - 4 = \mathbf{116}$
   (b) $\underline{6 \times 2} + \underline{8 \div 2} \times 4 = 12 + \underline{4 \times 4} = 12 + 16 = \mathbf{28}$
   (c) 71.2
   (d) 0.056

2. 5.629

3. 0.01

4. 3.75 $(3.07 \approx 3, 4.52 \approx 5, 4.99 \approx 5)$

5. 64 $(4 \times 16 = 64)$

6. 16.3

7. 2.44

8. $\frac{3}{5} < \frac{3}{4}$; same numerator, larger denominator.

   $\frac{3}{5}$ is $\frac{2}{5}$ from 1; $\frac{5}{7}$ is $\frac{2}{7}$ from 1, so $\frac{5}{7} > \frac{3}{5}$

   $\frac{3}{4} = \frac{21}{28}$; $\frac{5}{7} = \frac{20}{28}$; so $\frac{5}{7} < \frac{3}{4}$ Order: $\frac{3}{5}, \frac{5}{7}, \frac{3}{4}$

9. $2\frac{1}{3}$ h $= (2 \times 60) + (\frac{1}{3} \times 60) = 120 + 20 = \mathbf{140\ min}$

10. $1.2\ \ell = 1.2 \times 1000\ ml = \mathbf{1200\ ml}$

11. Total number of children: $30 + 10 = 40$

    $\frac{10}{40} \times 100\% = \mathbf{25\%}$

12. $7\% = \frac{7}{100} = \mathbf{0.07}$

13. $36\% = \frac{36}{100} = \frac{9}{25}$

14. Increase: 20% of $5 = \frac{1}{5} \times \$5 = \$1$

    (or: $10\% \to \$0.50$; $20\% \to \$1$)
    New fee = $\$5 + \$1 = \mathbf{\$6}$

15. $\angle DCB = 90° - 40° = 50°$ (right Δ)
    $\angle ACB = 50° - 35° = 15°$
    $\angle BAC = 90° - 15° = \mathbf{75°}$ (right Δ)

16. $\angle QRS = 180° - 74° = 106°$ (inside $\angle$'s //)
    $\angle TRS = 74°$ (iso. Δ)
    $\angle QRT = 106° - 74° = \mathbf{32°}$

17. $\angle x = 180° - 135° = \mathbf{45°}$ (inside $\angle$'s //)

18. Volume of one cube: $2\ cm \times 2\ cm \times 2\ cm = 8\ cm^3$
    There are 5 cubes. $5 \times 8\ cm^3 = \mathbf{40\ cm^3}$

19. (a) Take a base of the shaded triangle to be 6 in.
       The height from that base is 10 in.
       Area $= \frac{1}{2} \times$ (6 in. x 10 in.) = **30 in.**^{2}
    (b) Area of square: 10 in. x 10 in. = 100 in.^{2}
       Area of unshaded part: 100 in. − 30 in. = 70 in.^{2}
       Ratio of shaded to unshaded area:
       $30 : 70 = \mathbf{3 : 7}$

20. $\frac{416\ words}{8\ minutes} = \mathbf{52}$ **words per minute**

21. 2 months $\longrightarrow$ $300
    6 months $\longrightarrow$ $300 x 3 = $**900**

22. $5 $\longrightarrow$ 1 h
    $35 $\longrightarrow$ 1 h x 7 = **7 h**

23. (a) $3.50 x 3 = **$10.50**
    (b) 2 hours Tuesday: $7
       $22 − $7 = $15
       $15 ÷ $5 per hour = **3 h**

24.

    Peter ends up with 2 x (David's original amount plus the amount received). He started with $200.
    Let the amount received be 1 unit.
    1 unit = $200 − $90 − $90 = **$20**
    Or: They are given the same amount so the difference stays the same. The original difference is $200 − $90 = $110.
    1 unit = $110 − $90 = $20
    They were both given $20.

25.

    $\square = 1$ unit
    11 units = $110
    1 unit = $110 ÷ 11 = $10
    4 units = $10 x 4 = **$40**
    She had $40 left.

Unit 12 – Data Analysis

Chapter 1 – Mean, Median and Mode

Objectives

♦ Find the mean, median, and mode of a set of data.
♦ Understand how mean, median, and mode differ in the information they provide.

Material

♦ Graph paper
♦ Data on high and low temperatures or other data

Vocabulary

♦ Mean
♦ Median
♦ Mode
♦ Range

Notes

In *Primary Mathematics* 4B students learned to find the median and mode of a set of data. In Unit 11 of *Primary Mathematics* 5B they learned to find the average of a set of data, which is also the mean. In this chapter your student will find the mean, median, and mode of the same set of data and examine what kind of information each of these types of summary data provide.

The **mean** is the same as average and is calculated by dividing the sum of all the values in a set of data by the total number of values.

The **median** is the middle value in a set of data. When there is an odd number of values the median is the value in the middle. When there is an even number of values the median is the average of the two middle values.

The **mode** is the value that occurs most frequently in a set of data. At this level your student will only deal with cases in which there are only one or two modes.

The **range** is a measure of spread rather than central tendency and is the difference between the highest value and the lowest value. Students have found the range of a set of data in earlier levels.

The mean is the most commonly used type of summary data because it is an easy way to even out irregularities in the data. Sometimes the median is preferred to the mean, because it is less sensitive to extreme values since half of the values are above the median and half of the values are below it. Mean and median are usually used to analyze numerical data whereas mode is usually used to look at categorical data. Categorical data deals with categories that are not ordered, such as age group, race, gender, favorite ice cream, etc. Some data can be looked at both numerically and categorically (e.g., the yearly salaries of the employees in a company).

It is up to you how much time you want to spend on this chapter and whether you want your student to collect data and analyze it. The purpose of this chapter is just to introduce the student to the three common measures of central tendency, i.e., measures that represent the "center" of distribution, and how to compute them. The concepts will be covered in more depth in *Primary Mathematics* 6B, including the usefulness of each type of measure and the effect of adding a new data value to each.

(1) Find mean, median, mode, and range

Discussion

Concept p. 122

Have your student look at the table listing the salaries of different employees and make observations. Tell him that sometimes we want to summarize the data rather than looking at each individual piece of data or value. Discuss the methods to find the mean, median, and mode.

The *mean* is the average, which your student found in the previous unit.

The *median* is the middle value. When there are two middle values, as in this set of data, the median is the average of the two middle values. Point out that another way to find the average of two data values, rather than adding the two together and then dividing by 2, is to divide the difference by 2 and add that to the first number.

The *mode* is the value that appears the most. In this case it is obviously $40,000. Sometimes there is more than one mode. If no data value occurs more than once, then there is no mode.

The *range* is the difference between the highest and lowest value. Ask your student to find the range for this set of data. It is $100,000 − $40,000 = $60,000.

Discuss what the different summary values tell us about the data.

The average can tell us how much a company spends on salaries, since the total salary is the average times the number of employees, but does not really tell us what most employees make; the average is skewed high because of Mr. Capazzi's salary. You could go on an interview for a job and hear that the average salary is pretty high, and then find out that most employees make less than half of that, with the owner or the CEO or a few others making hundreds of times as much as the rest of the employees.

The median salary gives somewhat of a better idea of what most employees make since it is not as affected by an outlier or extreme value at either end. If the range is known it would tell us whether most employees are paid towards the bottom range of salaries or towards the top range of salaries.

In this set of data the mode gives the best idea of what a new hire would make since most employees make the same amount, which is at the lower end. In general, though, the mode is the least useful summary data, particularly if there are only a few values the same. If everyone made a different amount there would be no mode, and just one person making the same amount as another would make that the mode, whether it is at the high or low end. Mode is more useful when the data consists of only a few choices, rather than numbers. For example, mode is useful in looking at polling data to find out which candidate is most popular, or favorite choice among a limited number of categories, such as favorite ice-cream flavor.

In some cases the mean is the most informative summary data. It gives a better idea of future performance. For example, in sports a batting average which takes into account all the times a player has been up at bat, both good times and bad times, is probably the best indicator of how well he will do the next time he is at bat. School grades are usually based on average scores. But what if a student takes five tests and gets a perfect score and then fails one test because he was sick on the day of the test? Some teachers will also take into account the median or the mode in determining a final grade.

Point out that advertisers can use summary data in misleading ways. For example, an advertisement for some service that is supposed to save money over a competitor's service could give the average savings if you use their service, but most people would actually save much less.

Point out that data is only as good as the methods used to collect the data. For example, the mode for data on which product is being bought is very high for a new breakfast cereal. A company could start stocking up on that cereal since it is being bought more often. However, it could start out as seeming to be popular because many people are sampling it, but end up not buying it again, so additional data needs to be collected on repeat buys.

Task 1, p. 123

This task reviews line plots as a way of presenting data. To find the median ages we can make a list of the ages in increasing order to find the middle value. We can also cross out values on both sides of the chart, one pair at a time, until we get to the middle value or values.

1. (a) $\dfrac{9+(6\times10)+(5\times11)+(4\times12)}{16} = \dfrac{172}{16} = \mathbf{10.75}$

(b) 9, 10, 10, 10, **10, 10, 10, 11, 11, 11, 11, 11, 12, 12, 12, 12**

(c) 11

(d) 10

Practice

Task 2, p. 123

You can also have your student find the mode and range of this data.

2. (a) 79.2

(b) 65.9

(c) 73°, 74°, 74°, 75°, 79°, 81°, 82°, 83°, 85°, 86°
Median: (79° + 81°) ÷ 2 = **80°**
Or 79° + (81° − 79°) ÷ 2 = 80°
Or just look at the data, 80° is obviously halfway between 79° and 81°.

(d) 62°, 62°, 63°, 64°, 65°, 66°, 68°, 69,° 70°, 70°
Median: **65.5°**

Activity

If your student has been collecting data on high and low temperatures as suggested in this guide at the start of the previous unit, you can ask her to find the mean, median, mode, and range of the high and low temperatures.

If your student is collecting data for any science experiments, you can ask him to find mean, median, mode, and range for the data.

Workbook

Exercise 1, pp. 116-117 (answers p. 137)

Reinforcement

Extra Practice, Unit 12, Exercise 1, pp. 249-250

Test

Tests, Unit 12, 1A and 1B, pp. 155-159

Chapter 2 – Histograms

Objectives

♦ Categorize data into intervals and display the data using a histogram.
♦ Analyze and interpret data from histograms.

Material

♦ Graph paper
♦ Data on high and low temperatures or other data

Vocabulary

♦ Histogram

Notes

In *Primary Mathematics* 3 and 4 students learned to display data in bar graphs and to analyze data from bar graphs. In this chapter your student will learn to use a specific type of bar graph called a histogram.

A histogram is a representation of the frequency of data values by means of rectangles whose widths represent class intervals and whose areas are proportional to the corresponding number of data points that fall within the interval. The intervals are shown on the horizontal axis and the number of data points in each interval is represented by the height of a rectangle located above the interval. For example, a histogram can show the number of days in a year the temperature falls within different 5 degree ranges, or the height of plants in an experiment within 5-centimeter ranges, or the number of items sold in a month. The data are usually first put into a frequency table which shows how often an item, a number, or a range of numbers occurs.

The example at the right is for the number of students who scored within intervals of 20 points on a test. If the score is 20 or more but less than 40, the data value falls within the second interval shown. If all the data values are whole numbers, we can say there are 20 data points in the second interval, the whole numbers from 20 to 39.

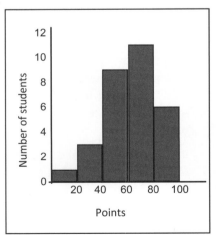

Histograms make it easy to see how frequently classes of data occur, the center and spread of the data, the skewness of the data, the presence of any outliers, and the mode of the data.

At this level it is not necessary to introduce your student to the term frequency table or frequency distribution, and the histograms will have only equal intervals.

(1) Understand histograms

Discussion

Concept pp. 124-125

The box at the top of p. 124 shows a collection of values in no particular order. The table shows the number of values that fall within specified ranges. It shows how frequently students scored within a range but does not show the actual values for each student. Discuss ways to conveniently transfer the data from the histogram to the table. One method is to make a table with the first column, and then write tally marks in each appropriate row as you systematically go through the data values. You may want to have your student do this and compare his results to those in the table.

(a) 24
(b) 10
(c) 8
(d) 80-89
(e) 60-69
(f) 80-89
(g) $\frac{4}{24}$ x 100% = $16\frac{2}{3}$%
(h) $\frac{2}{24}$ = $\frac{1}{12}$
(i) 8

The graph on the next page is a visual representation of this data. It is like a bar graph, except that each bar represents an interval and there are no spaces between the bars. This type of graph is called a *histogram*. Point out that the width of each bar represents a ten degree interval, but the individual data values used for each bar do not include any values for the right side of each bar. Since the data values are whole numbers, there are 10 data points in each interval. A value of 79.9, though, would go in the 70 up to but not including 80 interval.

As your student answers the questions on p. 125, you can discuss whether it is easier to use the histogram or the table to answer each question. For (a), (g), (h), and (i), which require exact numbers, it is probably easier to use the table. However, the graph is a good visual representation of the overall results and clearly shows the mode.

Practice

Tasks 1-2, pp. 126-127

For each of these assume the data values are all to the closest whole number, minute or centimeter.

Activity

If your student has been collecting high and low temperatures, you can have him create a frequency table and histogram of the information, choosing appropriate intervals. You can also have him collect other data to graph as a histogram.

Workbook

Exercise 2, pp. 118-121 (answers p. 137)

Reinforcement

Extra Practice, Unit 12, Exercise 2, pp. 251-252

Test

Tests, Unit 12, 2A and 2B, pp. 161-166

1. (a) 2 + 3 + 4 + 7 + 5 + 3
    = **24**
   (b) 50
   (c) 150-199
   (d) 3 + 4 + 7 = **14**
   (e) $\frac{8}{24}$ x 100% = $33\frac{1}{3}$%

2. (a) 2 + 8 + 14 + 9 + 3
    = **36**
   (b) 5 cm
   (c) 8
   (d) 150 cm to 154 cm
   (e) $\frac{9}{36}$ x 100% = **25%**
   (f) $\frac{9+3}{36}$ = $\frac{1}{3}$

Chapter 3 – Line Graphs

Objectives

♦ Represent data in a line graph.
♦ Analyze and interpret data in line graphs.

Material

♦ Graph paper
♦ Data on high and low temperatures or other data

Vocabulary

♦ Line graph
♦ Horizontal axis
♦ Vertical axis

Notes

In *Primary Mathematics* 4 students learned to interpret line graphs as another form of data representation. This is reviewed here.

The **line graphs** in this chapter display data collected over a period of time or other measure. Points are plotted similar to how they are plotted on a coordinate graph, with the independent variable on the x-axis, or **horizontal axis**, and the dependent variable on the y-axis, or **vertical axis**.

The independent variable is the value that does not depend on anything. In the example at the right for the height of a plant over time, the independent variable is the number of days. The dependent variable is the data value collected at a specific value for the independent variable. In the example at the right the dependent variable is the height of the plant. The height "depends" on the day it was measured.

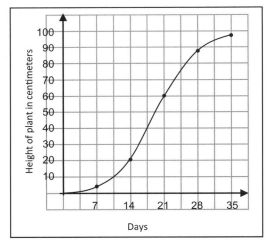

For visual convenience the data points are connected by lines drawn between successive data points. However, the points along the line between the data points are not valid data points. The dots indicate which points are actual measurements.

In histograms the data are grouped into intervals and the individual data points are not used, instead the number of data points in the interval are used. Histograms, as well as other bar graphs, are useful for seeing the mode and range of data. In line graphs the actual data points are used. Line graphs do not allow you to see the mode, but are useful for visualizing trends in the data. From a line graph, for example, we can immediately see the increase or decrease, over time, in the value of the measured quantity. In the example above the plant grew slowly for the first 2 weeks, and then more quickly for another two weeks, and then the rate of growth slowed again.

(1) Understand line graphs

Discussion

Concept pp. 128

> (a) Days
> (b) Temperatures
> (c) There is an overall increase in both the high and the low temperatures.

Tell your student that this kind of graph is called a *line graph* since the data points are connected with a line. The sides of the graph giving the scale are called the *axes* (singular: axis). The horizontal axis is for number of days, and the vertical axis for temperature.

Relate the information on the graph to the data in the table and how the data was plotted. For the data for high temperature at Day 1 we go up from 1 on the horizontal axis, and place the points at the appropriate place corresponding to the vertical axis. Depending on the graphing paper and how many squares are used for each unit, we sometimes have to estimate the placement of the data point. Usually, we draw a dot for the data points (the dots are not shown in the first printing of the textbook, but they are in Task 1 on the next page). After plotting all the points, we connect them with a line.

Point out that each data value is a temperature measured on a specific day. The value for the temperature is thus dependent on the day. The day is not dependent on anything, and is thus independent. In general the horizontal axis is used for independent values and the vertical axis for dependent values. The scale on the vertical axis needs to accommodate both the lowest and highest value. In this case, the highest value is 86°, so the vertical axis goes from 0° to 90°. Point out that as with bar graphs we don't label each point on the axis; in this case every 10 points are labeled. Although on these graphs the scales on both axes start at 0, they don't necessarily have to; the vertical axis could only go from 50° to 90°. Generally it is better to start at 0 when possible. The axes should also always be labeled with what they represent and any units (e.g., Day, Temperature in °F)

Ask your student what kind of information can be seen in a line graph. In this example it is easy to see how the temperature rises over a period of days.

Ask your student if a histogram would be a good way to display this data. If we had data for a full year and wanted to know how many days the temperature was in different ranges and wanted to find which range of data was the mode then a histogram could be useful, but generally data collected over time is better represented with a line graph in order to see the trends over time.

Practice

Tasks 1-2, pp. 129-130

> 1. (a) It increases
>   (b) 40 in.
>   (c) 7 years
>   (d) 9 and 10 years
>
> 2. (a) It decreases
>   (b) $5500
>   (c) No

Activity

If your student has been collecting data over time for your location, you can help her plot this data. Or you can provide data that has been collected over time, such as data from the internet on temperatures or humidity in your city, or other type of data pertinent to her current studies or interest.

To graph the data your student must first determine the size he wants to make the graph on the graph paper and then the scale for both axes; that is, what each unit on the graph paper represents. This involves determining the range that he wants to include (for example, 0 to 90), counting the number of squares on the graph paper within the dimensions of how large the graph should be, and then assigning a reasonable scale, such as 5 for each square. He then needs to draw the axes and label them, plot the points, and connect the points with straight lines.

Drawing a good graph and determining the scale takes some effort. Doing graphs by hand and plotting the points manually is good practice and helps students understand graphs better, even if they will eventually use computer software instead. Experience graphing can be obtained outside of "math" such as with science projects.

You can also teach your student to use graphing software such as the graphing component of spreadsheet types of program. Spreadsheet programs can make a variety of graphs, though histograms are more complicated to do and will show bars with spaces between them.

Workbook

Exercise 3, p. 122 (answers p. 137)

Reinforcement

Extra Practice, Unit 12, Exercise 3, pp. 253-254

Test

Tests, Unit 12, 3A and 3B, pp. 167-174

Workbook

Exercise 1, pp. 116-117

1. (a) 9, 9, 9, 9, 10, 10, 10, 11, 11, 12, 13, 13
   (b) 9
   (c) 10
   (d) (4 x 9) + (3 x 10) + (2 x 11) + 12 + (2 x 13) = 126
   126 ÷ 12 = **10.5**

2. (a) $3000, $3000, $3000, $4000, $6000
   (b) (3 x $3000) + $4000 + $6000 = $19,000
   $19,000 ÷ 5 = **$3800**
   (c) $3000
   (d) $3000

3. (a) 3.2, 3.4, 3.6, 4, 4.2, 4.5, 5.1, 5.4
   Median: Average of 4 and 4.2 = **4.1**
   (b) 3.2 + 3.4 + 3.6 + 4 + 4.2 + 4.5 + 5.1 + 5.4 = 33.4
   33.4 ÷ 8 = **4.175**

Exercise 2, pp. 118-121

1. (a)

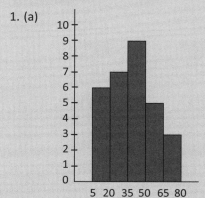

   (b) 30
   (c) 15 lb
   (d) 7
   (e) **35** lbs to **49** lbs
   (f) **65** lbs to **79** lbs
   (g) **35** lbs to **49** lbs
   (h) $\frac{9}{30}$ x 100% = **30%**
   (i) $\frac{3}{30} = \frac{1}{10}$

2. (a)

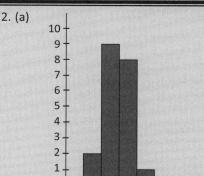

   (b) 20
   (c) 5
   (d) 9
   (e) **10** foul shots to **14** foul shots
   (f) **20** foul shots to **24** foul shots
   (g) $\frac{8}{20}$ x 100% = **40%**
   (h) $\frac{2}{20} = \frac{1}{10}$

Exercise 3, p. 122

1. (a)

Age	Average weekly income
14	$10
15	$17
16	$35
17	$59 (rounded)
18	$101

   (b)

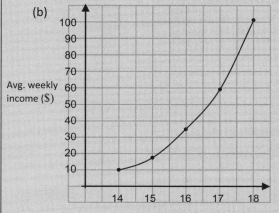

Avg. weekly income ($)

Age of students

   (c) The older the students were, the higher their income.

Chapter 4 – Pie Charts

Objectives

♦ Analyze and interpret data in circle graphs or pie charts.

Material

♦ Appendix p. a18

Vocabulary

♦ Pie chart
♦ Circle graph

Notes

Pie charts are used to show classes or groups of data in proportion to the whole data set. The entire pie represents all the data, while each slice represents a different class or group within the whole. Pie charts are also called circle graphs.

Three forms of pie charts are introduced in this unit: ones that display numbers (e.g., p. 132 in the textbook), ones that display fractions (e.g., p. 133), and ones that display percentages (e.g., p. 134).

Generally, pie charts would include either the values for all categories, often as a percentage, or no values if the focus is only on the relative size of each category. Actual data values would be in a table, and the values from the table used for computation. This curriculum takes advantage of pie charts to review converting between actual numbers, fractions, and percentages, as well as incorporating circles and degrees. The pie charts in this chapter include the values, fraction, or percentages of some but not all of the categories, and uses the little square symbolizing a right angle to show the 90° angle of a quarter of the circle. Assume that a straight line that looks like it goes through the center is a diameter and divides the circle in half.

Pie charts can be constructed by converting the fraction or percentage of the whole to a fraction or percentage of the total number of degrees in a circle and using the degrees to determine the angle at the center for each "piece" of the pie. Students are not required to draw pie charts, but you might want to include some experience with creating pie charts since it can be useful for reports in other subjects.

(1) Understand pie charts

Discussion

Concept p. 131

Have your student calculate the fraction for each category of T-shirts.

Tell your student that the circle of the pie chart or circle graph represents the total, 36 T-shirts. Each "slice" of the pie represents the same fraction of the total for each type of T-shirt. Although a pie chart tells which category is the mode, it does not tell us anything about the range or actual data values unless those are included. This type of graph is a good way to visually compare the sizes of the different categories to each other and to the whole. It is obvious from the chart that the store sells mostly size M T-shirts, and it should make sure that half its stock of T-shirts is size M.

Ask your student for ideas on how a pie chart could be constructed. Unless we have a pre-drawn circle with equal divisions around the edge, we would have to use degrees. To construct the graph we convert the fractions into degrees. 360° is the whole. One half of the T-shirts is then half of a turn around a circle, or 180°, and one fourth a quarter turn, or 90°. Ask your student to find the number of degrees that would be used to find the angle for a sixth and a twelfth of the total.

Practice

Tasks 1-2, p. 132

Activity

Ask your student to determine the degrees that would be used to construct the chart for Task 1.

Workbook

Exercise 4, pp. 123-126 (answers p. 142)

S shirts: $\frac{9}{36} = \frac{1}{4}$

M shirts: $\frac{18}{36} = \frac{1}{2}$

L shirts: $\frac{6}{36} = \frac{1}{6}$

XL shirts: $\frac{3}{36} = \frac{1}{12}$

S shirts: $\frac{1}{4}$ x 360° = 90°

M shirts: $\frac{1}{2}$ x 360° = 180°

L shirts: $\frac{1}{6}$ x 360° = 60°

XL shirts: $\frac{1}{12}$ x 360° = 30°

---

1. (a) Plastic chairs
   (b) 200 − 80 − 30 − 40 = **50**
   (c) $\frac{80}{200} = \frac{2}{5}$
   (d) **2** times as many

2. (a) $\frac{1}{4}$ (90° angle)
   (b) $3000 x 4 = **$12,000**
   (c) $12,000 − (2 x $3000) − $4800
      = **$1200**
   (d) 4800 : 3000 = **8 : 5**

Plastic chairs: $\frac{80}{200}$ x 360° = 144°

Wooden chairs: $\frac{50}{200}$ x 360° = 90°

Wicker chairs: $\frac{40}{200}$ x 360° = 72°

Wooden chairs: $\frac{30}{200}$ x 360° = 54°

(2) Interpret pie charts with fractions or degrees

Discussion

Tasks 3-4, p. 133

Task 3: In this pie chart actual values are not given, just the fraction of the total for most of the categories. If we know the total, we can find the values for the categories.

Task 4: One line divides the circle in half; the line is a diameter and goes through the center of the circle.

Practice

Appendix p. a18

The chart shows approximate sources of electricity in the U.S. in 2005. Have your student answer the questions. Answers are given at the right.

Workbook

Exercise 5, pp. 127-129 (answers p. 142)

---

3. (a) Toast

(b) $1 - \dfrac{3}{5} - \dfrac{1}{4} - \dfrac{1}{10} = \dfrac{20}{20} - \dfrac{12}{20} - \dfrac{5}{20} - \dfrac{2}{20} = \dfrac{\mathbf{1}}{\mathbf{20}}$

(c) $\dfrac{3}{5} \times 40 = \mathbf{24}$

(d) $\dfrac{1}{4} = \mathbf{25\%}$

4. (a) $\dfrac{1}{2}$

(b) $\dfrac{1}{4} = \mathbf{25\%}$

(c) $\dfrac{1}{4} - \dfrac{1}{8} = \dfrac{\mathbf{1}}{\mathbf{8}}$

(d) $\dfrac{1}{2} \longrightarrow 1200$

$\dfrac{1}{4} \longrightarrow 1200 \div 2 = \mathbf{600}$

(a) Coal

(b) $\dfrac{72}{360} = \dfrac{1}{5}$

$\dfrac{1}{5}$ of electricity comes from natural gas.

(c) $360 - 178 - 72 - 66 - 24 = 20$

$20 \div 2 = 10$

$\dfrac{10}{360} = \dfrac{1}{36}$

$\dfrac{1}{36}$ of electricity comes from petroleum.

(d) $\dfrac{66}{360} \times 100\% = 18\dfrac{1}{3}\%$

$18\dfrac{1}{3}\%$ of the electricity comes from nuclear energy.

(e) $\dfrac{24}{360} \times 1530 = 102$

102 billion kilowatt hours of electricity comes from hydroelectric power.

(3) Interpret pie charts with percentages

Discussion

Tasks 5-6, p. 134

Tell your student that pie charts showing percentages are more common than those that show actual amounts or fractions.

Task 6: The vertical line divides the circle into half. So the amount of money spent on pants and dresses is 50%, and the amount spent on shirts and shorts is also 50%.

Activity

Ask your student to find the degrees for each percentage value in Task 5.

You may want to guide your student in drawing a pie chart for this data or for other data using a compass to draw the circle, or you can use the circle on appendix p. a9, and a protractor. You may also want to teach your student to use computer software to generate pie charts.

Workbook

Exercise 6, pp. 130-133 (answers p. 142)

Reinforcement

Extra Practice, Unit 12, Exercise 4, pp. 255-260

Test

Tests, Unit 12, 4A and 4B, pp. 175-179

---

5. (a) Swimming

(b) $100\% - 30\% - 20\% - 35\% = \mathbf{15\%}$

(c) $\dfrac{30}{100} \times 200 = \mathbf{60}$

(d) $\dfrac{35}{100} = \dfrac{\mathbf{7}}{\mathbf{20}}$

6. (a) Shirts

(b) $50\% - 15\% = \mathbf{35\%}$

(c) $50\% - 20\% = \mathbf{30\%}$

Swimming: $\dfrac{35}{100} \times 360° = 126°$

Baseball: $\dfrac{30}{100} \times 360° = 108°$

Basketball: $\dfrac{15}{100} \times 360° = 54°$

Track and Field: $\dfrac{20}{100} \times 360° = 72°$

Workbook

Exercise 4, pp. 123-126

1. (a) 16
   (b) 12
   (c) 16 + 12 + 4 + 8 = **40**
   (d) $\dfrac{8}{40} = \dfrac{1}{5}$
   (e) $\dfrac{4}{40} = \dfrac{1}{10}$

2. (a) Cars
   (b) 2000
   (c) 3200 + 800 + 2000 + 4000 = **10,000**
   (d) $\dfrac{800}{10,000} = \dfrac{2}{25}$
   (e) $\dfrac{3200}{10,000} = \dfrac{8}{25}$

3. (a) 150 + 100 = **250**
   (b) Chicken + Egg = Tuna + Ham
   Number of chicken sandwiches:
   250 − 50 = **200**
   (c) 150 + 200 + 50 + 100 = **500**
   (d) $\dfrac{100}{500} \times 100\% = \textbf{20\%}$
   (e) $\dfrac{150}{50} = \textbf{3}$

4. (a) Entertainment is $\dfrac{1}{4}$ of his allowance.
   Transport and Savings are together $\dfrac{1}{4}$. Each
   is the same, so Savings is $\dfrac{1}{8}$ of his allowance.
   (b) $\dfrac{1}{8} \longrightarrow \$5$
   $1 \longrightarrow \$5 \times 8 = \$\textbf{40}$
   (c) $\dfrac{1}{2}$
   (d) $\dfrac{1}{4} = \textbf{25\%}$
   (e) $\dfrac{1}{8} : \dfrac{1}{2} = \dfrac{1}{8} : \dfrac{4}{8} = \textbf{1 : 4}$

Exercise 5, pp. 127-129

1. Motorcycles: $\dfrac{40}{200} = \dfrac{1}{5}$
   Cars: $\dfrac{80}{200} = \dfrac{2}{5}$
   Trucks: $\dfrac{30}{200} = \dfrac{3}{20}$

2. (a) $\dfrac{3}{8}$
   (b) $\dfrac{1}{4}$ (90° angle)
   (c) $1 - \dfrac{1}{4} - \dfrac{1}{4} - \dfrac{3}{8} = \dfrac{1}{8}$
   (d) $\dfrac{3}{8} \times 400 = \textbf{150}$
   (e) $\dfrac{1}{4} \times 400 = \textbf{100}$

3. (a) $\dfrac{1}{2}$
   (b) $\dfrac{1}{2} - \dfrac{1}{6} - \dfrac{1}{9} = \dfrac{9}{18} - \dfrac{3}{18} - \dfrac{2}{18} = \dfrac{4}{18} = \dfrac{2}{9}$
   (c) $\dfrac{1}{9} \times 180 = \textbf{20}$
   (d) $\dfrac{1}{6} \times 180 = \textbf{30}$
   (e) 30 − 20 = **10**

Exercise 6, pp. 130-133

1. Total = \$100 + \$150 + \$400 + \$350 = \$1000
   Fan: $\dfrac{100}{1000} = \textbf{10\%}$ Oven: $\dfrac{350}{1000} = \textbf{35\%}$
   Vacuum cleaner: $\dfrac{150}{1000} = \textbf{15\%}$

2. (a) 15%
   (b) 25%
   (c) 100% − 15% − 20% − 25% = **40%**
   (d) $\dfrac{15}{100} \times 80 = \textbf{12}$
   (e) $\dfrac{20}{100} \times 80 = \textbf{16}$

3. (a) 4
   (b) 50% − 25% − 15% = **10%**
   (c) $\dfrac{15}{100} \times 40 \text{ kg} = \textbf{6 kg}$
   (d) $\dfrac{1}{2} \times 40 \text{ kg} = \textbf{20 kg}$
   (e) $\dfrac{1}{4} \times 40 \text{ kg} = \textbf{10 kg}$

4. (a) $35\% = \dfrac{35}{100} = \dfrac{7}{20}$
   (b) 25%
   (c) 100% − 35% − 25% − 25% = **15%**
   (d) 25% → 30
   100% → 30 × 4 = **120**
   (e) $\dfrac{35}{100} \times 120 = \textbf{42}$

Review 12

Review

Review 12, pp. 135-139

Workbook

Review 12, pp. 134-138
(answers p. 145)

Tests

Tests, Units 1-12,
Cumulative Tests A and B,
pp. 181-187

1. (a) $60 \div \underline{(14-4)} \times 3$
$= \underline{60 \div 10} \times 3$
$= 6 \times 3$
$= \textbf{18}$

   (b) $50 - \underline{8 \times 2} + \underline{16 \div 8}$
$= 50 - 16 + 2$
$= \textbf{36}$

2. (a) $2\frac{1}{3}$ (b) 63

   (c) $\frac{1}{4}$ (d) $\frac{4}{27}$

3. (a) 3.79 (b) 0.867

4. $2.045 = 2\frac{45}{1000} = 2\frac{9}{200}$

5. (a) 4.0 (b) 7.64

6. 3000 km

7. Total of the four numbers: $82 \times 4 = 328$
   Total of three of the numbers: $63 + 74 + 85 = 222$
   Missing number: $328 - 222 = \textbf{106}$

8. $20 : 12 : 56 = 5 \times 4 : 3 \times 4 : 14 \times 4 = \textbf{5 : 3 : 14}$

9. (a) $\frac{20}{25} \times 100\% = \textbf{80\%}$

   (b) $\frac{90}{200} \times 100\% = \textbf{45\%}$

10. $48\% = \frac{48}{100} = \frac{\textbf{12}}{\textbf{25}}$

11. $2\frac{3}{5}$ hr $= (2 \times 60 \text{ min}) + (\frac{3}{5} \times 60)$ min $= 120$ min $+ 36$ min $= \textbf{156 min}$

12. $\frac{1}{2} \times 1.5$ kg $= \textbf{0.75 kg}$

13.

Each plate is \$1.20 more than a spoon. If \$1.20 were removed from the price of each of the 5 plates, then the resulting price would be the cost of 5 spoons. Total cost would be the cost of $8 + 5 = 13$ spoons.
13 units $= \$16.40 - (5 \times \$1.20) = \$10.40$
1 unit $= \$10.40 \div 13 = \textbf{\$0.80}$
Each spoon cost \$0.80.

14. Money from chickens: $\$3.50 \times 200 = \$700 = $ cost of 30 turkeys

   1 turkey cost $\$\frac{700}{30} = \textbf{\$23.33}$ (rounded to the nearest cent)

(Continued next page)

15. Total weight of both boys: 48 kg x 2 = 96 kg

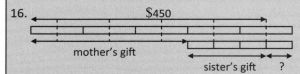

If 6 kg is added to the weight, both would weigh the same as the heavier boy.
2 units = 96 kg + 6 = 102 kg
1 unit = 102 kg ÷ 2 = **51 kg**
The heavier boy weighs 51 kg.

16.

$450

mother's gift

sister's gift   ?

(a) To get 3 fourths of the remaining 2 fifths, divide each fifth into half, giving 10 units.

He spent $\frac{3}{10}$ of his money on his sister's present.

(b) 9 units = $450
1 unit = **$50**
He had $50 left.

17. 50 words → 1 min

300 words → $\frac{1}{50}$ x 300 min = 6 min

4 pages will take 6 min x 4 = **24 min**.

18. Rate for first 1.5 km: $2.40
4 km − 1.5 km = 2.5 km = 2500 m = 25 x 100 m
He traveled 25 additional 100 m beyond the first 1.5 km.
Fare for additional distance: 25 x $0.10 = $2.50
Total fare: $2.40 + $2.50 = **$4.90**

19. men

women

12

2 units = 12

12 units = $\frac{12}{2}$ x 12 = **72**

There are 72 members altogether.

20. Percentage of girls: 100% − 60% = 40%
There are 20% of total more boys than girls.

20% of 1800 = $\frac{20}{100}$ x 1800 = **360**
There are 360 more boys than girls.

21. (a) ∠ACB = 58°   (iso. Δ)
    ∠y = 58° − 26° = **32°**   (ext. ∠ of Δ)
(b) ∠CDB = 180° − 145° = 35° (∠'s on st. line)
    ∠y = 62° + 35° = **97°**   (ext. ∠ of Δ)

22. 15 in. x 16 in. = **240 in.**^{2}

23. $\frac{12,000 \text{ cm}^2}{40 \text{ cm} \times 25 \text{ cm}}$ = **12 cm**

The water is 12 cm high.

24. 60 cm x 20 cm x 30 cm = 36,000 cm^3 = 36 ℓ
36 ℓ − 28 ℓ = **8 ℓ**
8 liters more water is needed.

25. (a) Total boys: 23 + 18 + 17 + 20 = 78
    Total girls: 15 + 20 + 19 + 18 = 72
    78 − 72 = **6**
    There are 6 more boys than girls.
(b) Total students in 5th grade: 78 + 72 = 150

    $\frac{78}{150}$ x 100% = **52%**

    52% of the students are boys.

26. (a) (3 x 9) + (7 x 10) + (7 x 11) + (3 x 12) = 210
    210 ÷ 20 = **10.5**
(b) 10.5 (average of 10 and 11)
(c) 10 and 11 (2 modes)

27. (a) 200 ℓ + 250 ℓ + 150 ℓ + 300 ℓ + 200 ℓ + 100 ℓ
    = 1200 ℓ
    1200 ℓ ÷ 6 = **200 ℓ**
    He bought an average of 200 ℓ of gas per month.
(b) 250 ℓ − 150 ℓ = 100 ℓ
    100 x $1.15 = **$115**
    He spent $115 less on gas in March than Feb.

28. (a) Thursday

(b) $\frac{2}{40}$ x 100% = **5%**

(c) 40 + 37 + 36 + 34 + 38 = 185
    185 ÷ 5 = **37 students per day**

29. (a) 50% − 20% − 12% = **18%**
    18% of the students liked the school band.

(b) 20% = $\frac{1}{5}$ $\frac{1}{5}$ of the students liked swimming.

(c) 12% → 18

    100% → $\frac{18}{12}$ x 100 = **150**

    There were 150 students in the group.

30. (a) 1 fourth is 20, so half is **40**.
    40 students chose the Art club.
(b) **25%** of the students chose Chess club.
(c) 20 − 12 = **8**
    8 students chose the Drama Club.
(d) 20 x 4 = **80**
    80 students were in the group.

Workbook

Review 12, pp. 134-138

1. (a) 40.66
   (b) 23.29
   (c) 284.4
   (d) 573.75

2. (a) 9.059   (b) 8.603

3. (a) 18   (b) 66

4. 30

5. 2 x 7 x 11

6. (a) 2 yd 2 ft x 5 = 10 yd 10 ft = **13** yd **1** ft
   (b) 6 gal 3 qt x 6 = 36 gal 18 qt = **40** gal **2** qt
   (c) 5 qt 2 c x 7 = 35 qt 14 c = **38** qt **2** c
   (d) 6 qt 1 pt x 4 = 24 qt 4 pt = **26** qt **0** pt
   (e) 5 lb 12 oz ÷ 4 = 1 lb + (28 oz ÷ 4) = **1** lb **7** oz
   (f) 3 ft 8 in. ÷ 11 = 44 in. ÷ 11 = **0** ft **4** in.
   (g) 7 gal 2 qt ÷ 6 = 1 gal + (6 qt ÷ 6) = **1** gal **1** qt

7. (a) 9   (b) 30

8. (a) 3   (b) $8\frac{2}{3}$ (just double $4\frac{1}{3}$)
   (c) 5   (since the answer is the numerator, use a
      number that will get rid of the denominator.)
   (d) 24 ($\frac{1}{3}$ of 48 is 16, so $\frac{2}{3}$ of half that is 16.)

9. 60°   (iso. Δ)

10. (a) 2.2 mi + (2.2 mi − 0.7 mi) = **3.7 mi**
    (b) (1.5 mi + 2.2 mi) ÷ 2 = 3.7 mi ÷ 2 = **1.85 mi**

11. Weight of 5 bars: 3.4 lb − 1.1 lb = 2.3 lb
    Weight of 1 bar: 2.3 lb ÷ 5 = **0.46 lb**

12. ($\frac{1}{2}$ x $3.50) + (2 x $1.95) = $1.75 + $3.90 = **$5.65**

13. $\frac{5}{6}$ → 10 gal   Or: 10 gal ÷ $\frac{5}{6}$ = 12 gal
    $\frac{1}{6}$ → $\frac{10}{5}$ gal = 2 gal
    $\frac{6}{6}$ → $\frac{10}{5}$ gal x 6 = **12 gal**

14. 253 in. − (10 x 11 in.) = 253 in. − 110 in. = **143 in.**

15. 3 ft 7 in. ÷ 4 = 43 in. ÷ 4 = 10.75 in. ≈ **11 in.**

16. 50 words → 1 min
    1 word → $\frac{1}{50}$ min
    1800 words → $\frac{1}{50}$ x 1800 = **36 min**

17. 400 ml → 1 min
    7200 ml → $\frac{1}{400}$ x 7200 = **18 min**

18. $800 − (15% x $800) = $800 − $120 = **$680**

19. 6 x 8 cm x 8 cm = **384 cm^2**

20.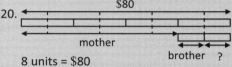
    8 units = $80
    1 unit = **$10**
    Or: $\frac{1}{2}$ x $\frac{1}{4}$ x $80 = $10

21. Height of top triangle: 18 cm − 10 cm = 8 cm
    ($\frac{1}{2}$ x 10 cm x 8 cm) + ($\frac{1}{2}$ x 6 cm x 10 cm) + 100 cm^2
    = 40 cm^2 + 30 cm^2 + 100 cm^2
    = **170 cm^2**

22. (a) 25%
    (b) 50% − 10% − 25% = **15%**
    (c) $200 x 2 = **$400**

23. (a)

Score	Tally	Number
0-19	/	1
20-39	///	3
40-59	*LHT* ////	9
60-79	*LHT LHT* /	11
80-99	*LHT* /	6

(b)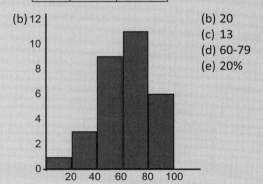

(b) 20
(c) 13
(d) 60-79
(e) 20%

Unit 13 – Algebra

Chapter 1 – Algebraic Expressions

Objectives

- Use letters to represent unknown numbers.
- Write algebraic expressions.
- Use exponents in algebraic expressions.
- Evaluate algebraic expressions using substitution.
- Simplify algebraic expressions in one unknown.

Material

- Game tokens and unit cubes or other items
- Mental Math 13 (appendix)

Vocabulary

- Algebraic expression
- Express in terms of
- Substitute
- Evaluate
- Coefficient
- Constant
- Simplify

Notes

Arithmetic expressions can involve addition, subtraction, multiplication, and division. In this chapter your student will learn about arithmetic expressions which involve an unknown value, called **algebraic expressions**. In algebraic expressions, the unknown value is represented by a letter such as n, and we are free to assign any value to the letter. For example, if we **substitute** a value of 3 for n in the expression $n + 2$, then the expression is **evaluated** as $3 + 2 = 5$, and the expression is then equal to 5. If instead we substitute a value of 5 for n, the expression is evaluated as $5 + 2 = 7$.

In later grades your student will learn that letters which represent an unknown are called variables. Variables can be assigned a range of values. Letters as variables can also stand for constants, parameters, or can be used to name objects. The complete definition involved in the concept of variable is more complex than students need at this level, and the term variable is not used in *Primary Mathematics*. For this unit students should think of the letter as standing for a number in an arithmetic expression. The expression can be evaluated by assigning a specific value to the letter. The letter then becomes that number.

Students have already been informally exposed to the idea of finding the unknown value starting as early as *Primary Mathematics* 1A. At that level, the unknown was the missing part in a number bond, or an empty rectangle, $\square + 4 = 7$, or a blank line, _____ $+ 4 = 7$. In *Primary Mathematics* 4A, sometimes a letter was used, $n + 2 = 7$, where the letter clearly fulfilled the same function as a blank. Students solved for the unknown in the same way that they would if a blank were used, primarily by thinking in terms of part-whole and related equations. For example, $n + 4 = 7$ is missing a part and so can be solved by solving $7 - 2$.

In this chapter your student will only encounter algebraic *expressions*. In *Primary Mathematics 6*, students will learn more formally how to solve algebraic *equations* for the unknown.

In *Primary Mathematics 3A* students were introduced to diagrams in which an unknown quantity was represented by a bar labeled with a question mark. Now they can use a letter to represent the unknown value. You can help your student see the analogy between a letter representing an unknown and the familiar unknown in a bar model by labeling the bar whose value is presently unknown with a letter.

In a term containing both a number part and an unknown part, such as $2y$, the number part of the term, 2, is called a **coefficient**. In $5y$, the coefficient is 5. A term that contains only numbers (no unknowns) is called a **constant**. In $2y + 3$, the 3 is a constant. You can teach your student these words so that you can use them in teaching.

In this chapter your student will learn to simplify algebraic expressions by combining like terms. Like terms are those terms in an expression that have the exact same letters *and* exponents. $2y$ and $3y$ are like terms, $2a$ and $2b$ are not, and $2y$ and $3y^2$ are not. In this unit your student will only encounter expressions with constants or with like terms in one unknown; that is, they will not encounter expressions such as $3y + 4x - y + 10 + 7y$, which has two unknowns, x and y.

Addition and subtraction can be done in any order, as long as the plus or minus signs are kept with the number (otherwise, addition and subtraction must be done from left to right). In performing an arithmetic operation on expressions, we are simplifying the expression. For example:

$$3 - 10 + 8 = 3 + 8 - 10 = 1$$

Similarly, algebraic expressions can also be simplified by combining like terms together and constants together. We can do this by first grouping like terms. For example:

$$3y - 10y + 8y = 3y + 8y - 10y = y$$

or

$$4y - 5 - 2y + 10 = 4y - 2y + 10 - 5 = 2y + 5$$

The sign before each term must stay with its term. The fact that the "− 5" term can be moved after the "+ 10" term can cause confusion for some students. Students have learned that addition is commutative and can be done in any order, but that subtraction is not. In *Primary Mathematics* 5A students learned to add and subtract from left to right. This automatically kept the minus sign with the term that follows it.

Illustrate combining like terms concretely. Choose two types of uniform objects, one object to represent the unknown, such as a identical game tokens (e.g., pawns from a chess set), and another (smaller) object to represent ones (constants), such as the unit cubes from a base-10 set. Your student can set out markers and cubes for terms to be added, then take away some for terms to be subtracted. For example, for $4y - 5 - 2y + 10$, she can set out 4 markers for $4y$ and 10 cubes for 10, and then take away two of the markers and 5 cubes. The first term and the ones with "+" in front indicate what are added in first, and the terms with "−" in front indicate what is then taken out.

To illustrate combining like terms pictorially, you can draw bags and marbles as in the textbook. Each bag contains the same unknown number of marbles, and each marble stands for a one.

(1) Write simple algebraic expressions

Discussion

Concept p. 140

Discuss the table in the top half of p. 140. Your student should observe that Limei is 2 years older than Angela.

Draw the table for Limei's and Angela's ages similar to that in the textbook but include an equation in the second column. Ask your student if he sees a pattern in the equations. In each, 2 years are added to Angela's age.

Tell your student that if we don't know Angela's age at first, or if we want a general expression to show how Limei's age relates to Angela's age, we can use the letter n to stand for Angela's age. The letter n can be any number that makes sense within the context of this situation. Limei's age could then be written as $n + 2$. Add another line to the table to show this. Tell your student that $n + 2$ is called an *algebraic expression*. It contains an unknown number represented by n. When the value of n changes, the value of the entire expression, $n + 2$, changes accordingly. Show your student how the expression can be represented by a bar model. Since we don't know the bar length for n relative to the bar length for 2 until we assign a number to n, we can think of the bar for n as being elastic — it can stretch or shrink depending on the value for n.

If we say that n stands for Angela's age, then we can say that we are *expressing* Limei's age *in terms of n*, or in terms of Angela's age. If we are told what Angela's age is, we can *substitute* the number we are given for Angela's age for n, and then find Limei's age.

Tell your student that It does not matter what letter we choose to stand for the unknown. We could use a instead and write it as $a + 2$. Although we could write the expression as $2 + n$ it is customary to write the part of the expression with the letter for the unknown first.

Tasks 1-3, p. 141

Tell your student that she can always draw or imagine a bar model if she is unsure of how to write the expression. In Task 2, for example, we are told that Jim has $2 more than Travis, but we need to write an expression for how much Travis has in terms of the amount Jim has. Since Travis has less than Jim, we need to use subtraction. A bar model will make it more obvious that we need to use subtraction.

The parentheses simply means that the units apply to both values within the parentheses.

> When Angela was 12 years old, Limei was **14** years old.
> When Angela was 15 years old, Limei was **17** years old
> When $n = 20$, $n + 2 =$ **22**

Angela's age	Limei's age
6	$6 + 2 = 8$
7	$7 + 2 = 9$
8	$8 + 2 = 10$
9	$9 + 2 = 11$
10	$10 + 2 = 12$
n	$n + 2$

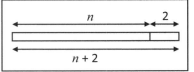

$n + 2$
$a + 2$
$2 + n$

1. (a) 13 years
   (b) $(x + 8)$ years
2. (a) $8
   (b) $(m - 2)$
3. (a) $(w - 5)$ kg
   (b) 3
      3 kg

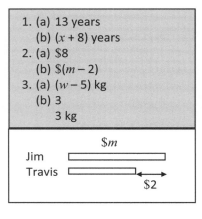

Task 4, p. 142

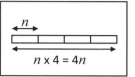

Task 4(a): Make sure your student sees the pattern — the second factor in the equation is the same as the number of packages. If we let *n* equal the number of packages, we can write a general expression $4 \times n$ for the total number of apples. Point out that in algebraic expressions, we can omit the multiplication sign and write $4 \times n$ as $4n$. We also write $n \times 4$ as $4n$. Ask him to draw a bar diagram to illustrate the expression. The unit is *n*.

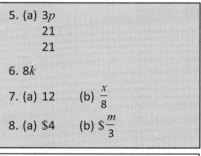

Task 4(b): Tell your student that when we *substitute* 8 in for *n*, we are *evaluating* the expression $4n$ when $n = 8$.

Tell your student to suppose we have three crates, each crate containing *n* bags, each bag containing 4 apples. We can write the total number of apples as $3 \times 4n$. In this expression, we cannot simply eliminate the multiplication symbol between the two numbers, as we can between the 4 and the unknown. Tell your student that sometimes, in other math books, the multiplication sign is replaced by a raised dot, with $3 \times 4n$ shown as $3 \bullet 4n$. Multiplication is also sometimes shown using parentheses. In most texts the letter being used for an unknown is italicized. If there is likely to be any confusion between x for multiplication and *x* for an unknown then don't use the letter *x* for the unknown.

Tell your student that in the term $4n$ the "4" part is called the *coefficient*. The coefficient is the part multiplied by the unknown. In the expression $n + 2$, *n* has a coefficient of 1, but we write $1n$ as simply *n*. The term 2 is called a *constant* since it does not change.

Tasks 5-8, pp. 142-143

Again, if your student is uncertain of what equation to use in the algebraic expression, she can draw a bar model. You can have her draw a bar model for Task 7. Tell her

that although we could write the expression $\dfrac{x}{8}$ with a

division symbol, the division symbol is not normally used in algebraic expressions; instead we write it as a fraction.

Since $\dfrac{x}{8}$ can be written as $\dfrac{1}{8}x$ the coefficient for this term

is $\dfrac{1}{8}$.

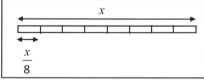

Practice

Task 9, p. 143

Workbook

Exercise 1, pp. 139-141 (answers p. 155)

Right sidebar boxes:

4. (a) $4n$
   (b) 32
   (c) 44

```
3 x 4n
3 ● 4n
(3)(4n)
```

5. (a) $3p$
   21
   21

6. $8k$

7. (a) 12   (b) $\dfrac{x}{8}$

8. (a) \$4   (b) \$$\dfrac{m}{3}$

9. (a) 10   (b) 16   (c) 9
   (d) 0   (e) 24   (f) 60
   (g) $\dfrac{3}{4}$ (h) 1   (i) $\dfrac{1}{2}$

(2) Write and evaluate algebraic expressions

Discussion

Task 10, p. 144

You may want to draw bags and circles as you discuss this task, or you can use concrete manipulatives where some object, such as a game piece, stands for the unknown amount x, and another object, such as a unit cube, stands for the ones. For bags, first draw 5 bags, say there are x marbles in each, and ask your student to write an expression for the total number of marbles. Add circles for 3 marbles and ask him to write a new expression. Then look at the picture in the book. Ask him to draw a bar diagram to illustrate the expression. It would be a part-whole diagram, with 5 equal units representing x for one part.

In evaluating the expression $5x + 3$ for $x = 10$ we follow order of operations and do the multiplication first.

Write the following problems and ask your student to write possible expressions to represent the given information.

⇒ Sam had 5 bags of marbles, with 10 marbles in each bag. He added 3 marbles to each bag. How many marbles does he now have?

⇒ Sam had 5 bags of marbles, with m marbles in each bag. He added 3 marbles to each bag. How many marbles does he now have?

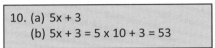

10. (a) $5x + 3$
    (b) $5x + 3 = 5 \times 10 + 3 = 53$

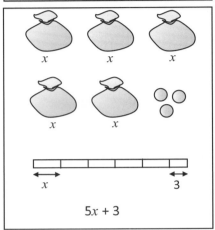

$(10 + 3) \times 5$
or $5 \times (10 + 3)$
or $5 \times 10 + 5 \times 3$

$(m + 3) \times 5$
or $5 \times (m + 3)$
or $5m + 5 \times 3 = 5m + 15$

Tasks 11-16, pp. 144-145

Task 12: You can ask your student to draw a bar model for this task. Even though we do not know how long the bar for y is compared to the bars for the amount he gave the two daughters, a model can be helpful in deriving an appropriate expression. Remind her that we write division as a fraction in algebraic expressions. Point out that by doing so it is easy to see that we need to subtract y from 50 before dividing.

Task 15: Since your student has learned to write exponents with whole numbers already, tell him that we can also write that $y^2 = 3^2$ when $y = 3$.

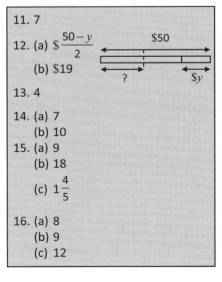

11. 7
12. (a) $\dfrac{50 - y}{2}$
    (b) $19
13. 4
14. (a) 7
    (b) 10
15. (a) 9
    (b) 18
    (c) $1\dfrac{4}{5}$
16. (a) 8
    (b) 9
    (c) 12

Practice

Task 17, p. 145

Activity

Students sometime confuse the exponent on measurements with the exponent on letters representing unknowns and think they should match.

Tell your student you have a cube that measures s cm on the side. Ask her to write an equation for its volume and include the units. Tell her that the correct notation is to write the exponent both for the letter (s) and the unit (cm).

Now, tell your student that you have a box with a square base, b cm on the side and 5 cm tall. Ask him to write an expression for the area of the base, and then for the volume of the box, using exponents and including the measurement units.

Point out that since we are finding volume, the volume is centimeter cubed, and we write an exponent of 3 for the cm. But the exponent of the unknown, b, in the expression is 2, not 3, because we are only multiplying 2 b's together. Tell your student not to expect that the exponent for the unknown will be the same as that for the measurement unit.

Workbook

Exercise 2, pp. 142-143 (answers p. 155)

17. (a) $6\frac{2}{3}$ (b) 23 (c) 7

    (d) $3\frac{2}{3}$ (e) $5\frac{1}{2}$ (f) 3

    (g) 47 (h) 130 (i) 120

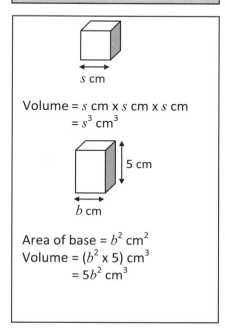

Volume $= s$ cm x s cm x s cm
$= s^3$ cm^3

Area of base $= b^2$ cm^2
Volume $= (b^2 \times 5)$ cm^3
$= 5b^2$ cm^3

(3) Simplify algebraic expressions

Discussion

Task 18, p. 146

Since each bag has the same number of marbles, whether green or red, we can add or subtract the number of each type of bag first and then multiply by the number of marbles in each bag rather than first finding the number of red and green marbles separately. Point out that we are applying the distributive property. You can use actual numbers to review this property if needed. You can also draw a bar model to illustrate this concept. All the units are x so they are all the same.

Tell your student that when there is an algebraic expression with more than one term with the same unknown, we can combine those terms and *simplify* the expression. That will make evaluating the expression simpler when we are given a value for the unknown.

Task 19, p. 147

Make sure your student understand that we apply the distributive property only to terms with the unknown. We simplify the constants separately. When we rearrange the terms, the operation stays with the term just like when we apply order of operations. For example, $5r + 3 - 2r + 3r = 5r - 2r + 3r + 3$; the subtraction symbol stays with the $2r$ since that is what we are subtracting. If necessary, illustrate each of these with manipulatives such as identical game tokens and unit cubes, first adding in tokens and cubes for all terms after a plus sign and then taking out tokens or cubes for terms following a minus sign.

Practice

Task 20, p. 147

Workbook

Exercise 3, pp. 144-145 (answers p. 155)

18. (a) $4x + 3x = (4 + 3)x = \mathbf{7x}$
    (b) $4x - 3x = (4 - 3)x = \mathbf{x}$

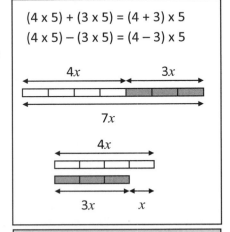

$(4 \times 5) + (3 \times 5) = (4 + 3) \times 5$
$(4 \times 5) - (3 \times 5) = (4 - 3) \times 5$

19. (a) $3r$
    (b) $6r$
    (c) $3r + 3$
    (d) $6r + 3$
    (e) $7k + 3$

20. (a) $9a$ (b) $3c$
    (c) $6k$ (d) $2x + 6$
    (e) $5m + 7$ (f) $7s + 10$
    (g) $5y + 3$ (h) $m + 1$
    (i) $6r$ (j) $4p + 2$
    (k) $6w + 13$ (l) $3h$

(4) Practice

Practice

Practice A, p. 148

Reinforcement

Extra Practice, Unit 13, Exercise 1, pp. 269-274

Mental Math 13

Enrichment

See next page of this guide.

Test

Tests, Unit 13, 1A and 1B, pp. 189-192

1. (a) $21 - y$
   $= 21 - 4$
   $= \mathbf{17}$

   (b) $y + 25$
   $= 4 + 25$
   $= \mathbf{29}$

   (c) $3y + 2$
   $= 3 \times 4 + 2$
   $= 12 + 2$
   $= \mathbf{14}$

   (d) $3y = 3 \times 4 = \mathbf{12}$

   (e) $\dfrac{y}{2} = \dfrac{4}{2} = \mathbf{2}$

   (f) $\dfrac{y}{16} = \dfrac{4}{16} = \dfrac{\mathbf{1}}{\mathbf{4}}$

   (g) $\dfrac{2y-5}{4} = \dfrac{2 \times 4 - 5}{4}$
   $= \dfrac{\mathbf{3}}{\mathbf{4}}$

   (h) $y^2 + 4 = 4^2 + 4$
   $= 16 + 4$
   $= \mathbf{20}$

   (i) $2y^2 = 2 \times 4 \times 4$
   $= \mathbf{32}$

   (j) $y^3 - 20$
   $= 4 \times 4 \times 4 - 20$
   $= 64 - 20$
   $= \mathbf{44}$

   (k) $\dfrac{3y}{2} = \dfrac{3 \times 4}{2}$
   $= \mathbf{6}$

   (l) $50 - 3y^2$
   $= 50 - 3 \times 4 \times 4$
   $= 50 - 48$
   $= \mathbf{2}$

2. (a) $x + x + x = \mathbf{3x}$
   (b) $3x + 4x = \mathbf{7x}$
   (c) $6p - 4p = \mathbf{2p}$

3. (a) $2p + 2p - p = \mathbf{3p}$
   (b) $4r - 2r + 3r = \mathbf{5r}$
   (c) $5f - f - 3f = \mathbf{f}$

4. (a) $3c - 3c + c = \mathbf{c}$
   (b) $5k + 7 - k$
   $= 5k - k + 7$
   $= \mathbf{4k + 7}$
   (c) $6n + 3 + n + 2$
   $= 6n + n + 3 + 2$
   $= \mathbf{7n + 5}$

5. (a) $7g - 2g + 2$
   $= \mathbf{5g + 2}$
   (b) $10x + 5 - 4x - 2$
   $= 10x - 4x + 5 - 2$
   $= \mathbf{6x + 3}$
   (c) $3h + 8 - 3h + 2$
   $= 3h - 3h + 8 + 2$
   $= \mathbf{10}$

6. (a) $\$(y + 1)$
   (b) $\$(8 + 1) = \mathbf{\$9}$

7. (a) $3x$ m
   (b) (3×9) m $= \mathbf{27\ m}$

8. (a) $3x + 4$ years
   (b) $3 \times 4 + 4 = 12 + 4 = \mathbf{16\ years}$

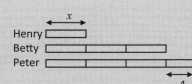

9. (a) $\dfrac{\$(50 - y)}{\$2}$
   (b) $\dfrac{\$(50 - 38)}{\$2} = \dfrac{12}{2} = 6$

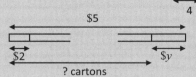

Enrichment

Have your student come up with original story problems to go with some algebraic expressions. You can use some of the ones in Practice 1A.

Ask your student to do the following number trick:

- Choose a number.
- Add 5.
- Double the result.
- Subtract 4.
- Divide the result by 2.
- Subtract the number you started with.

The result is 3. Have your student pick other starting numbers and see that the result will always be 3. Guide him through writing algebraic expressions for each step, using an unknown for the number picked. You can illustrate with markers and cubes or by drawings bags and marbles.

Pick a number.	n
Add 5.	$n + 5$
Double the result.	$2n + 10$ (note that both terms must be doubled)
Subtract 4.	$2n + 10 - 4 = 2n + 6$
Divide the result by 2. Use the distributive property; each term has to be divided by 2.	$\dfrac{2n+6}{2} = \dfrac{2(n+3)}{2} = n + 3$
Subtract the number you started with.	$n + 3 - n = 3$. The result is always 3.

Have your student try another number trick with several numbers, determine what the result will always be, then prove that the result will always be 1 using algebraic expressions:

- Pick a number.
- Add 3.
- Double the result.
- Subtract 4 from the result.
- Divide the result by 2.
- Subtract the original number.

Let your student come up with his own number trick.

This number trick is a bit more challenging to prove algebraically and involves two unknowns and the place value concept:

Pick a number between 0 and 9.	n
Double it.	$2n$
Add 5.	$2n + 5$
Multiply by 5.	$10n + 25$
Pick another number between 1 and 9 and add its value to the total.	$10n + 25 + m$
Subtract 25 from the total.	$10n + 25 + m - 25 = 10n + m$
The two digits of the result are the same as the two numbers.	$10n + m$ gives a 2 digit number with the first digit the tens and the second digit the ones.

Workbook

Exercise 1, pp. 139-141

1.

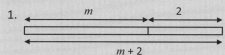

m 2

m + 2

(a) (*m* + 2) kg
(b) (*m* + 2) kg = (4 + 2) kg = **6 kg**
(c) (*m* + 2) kg = (6 + 2) kg = **8 kg**

2.

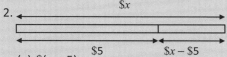

$x

$5 $x − $5

(a) $(*x* − 5)
(b) $(*x* − 5) = $(11 − 5) = **$6**
(c) $(*x* − 5) = $(15 − 5) = **$10**

3.

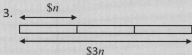

$n

$3n

(a) $3*n*
(b) $3*n* = 3 × $8 = **$24**
(c) $3*n* = 3 × $10 = **$30**

4. (a) $\dfrac{w}{4}$ cm

(b) $\dfrac{w}{4}$ cm $= \dfrac{592}{4}$ cm = **148 cm**

(c) $\dfrac{w}{4}$ cm $= \dfrac{608}{4}$ cm = **152 cm**

5. (a) $n + 7 = 15 + 7 =$ **22**   (b) $20 - n = 20 - 15 =$ **5**
   (c) $3n = 3 \times 15 =$ **45**   (d) $n + 5 = 15 + 5 =$ **20**

(e) $\dfrac{n}{5} = \dfrac{15}{5} =$ **3**   (f) $n - 3 = 15 - 3 =$ **12**

(g) $\dfrac{n}{3} = \dfrac{15}{3} =$ **5**   (h) $5n = 5 \times 15 =$ **75**

(i) $3 + n = 3 + 15 =$ **18**   (j) $\dfrac{n}{45} = \dfrac{15}{45} = \dfrac{1}{3}$

Exercise 2, pp. 142-143

1. mango   papaya

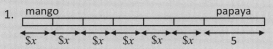

$x $x $x $x $x $x   5

(a) $(6*x* + 5)
(b) $(6*x* + 5) = $(6 x 2 + 5) = **$17**
(c) $(6*x* + 5) = $(6 x 3 + 5) = **$23**

2.

$40

bottle $y

(a) $ $\dfrac{40 - y}{6}$

(b) $ $\dfrac{40 - y}{6}$ = $ $\dfrac{40 - 10}{6}$ = $ $\dfrac{30}{6}$ = **$5**

(c) $ $\dfrac{40 - y}{6}$ = $ $\dfrac{40 - 1}{6}$ = $ $\dfrac{39}{6}$ = **$6.50**

3. (a) $\dfrac{5k}{3} = \dfrac{5 \times 6}{3} = \dfrac{30}{3} =$ **10**

(b) $\dfrac{15 - k}{3} = \dfrac{15 - 6}{3} = \dfrac{9}{3} =$ **3**

(c) $\dfrac{8 + k}{7} = \dfrac{8 + 6}{7} = \dfrac{14}{7} =$ **2**

(d) $10 - \dfrac{2k}{3} = 10 - \dfrac{2 \times 6}{3} = 10 - 4 =$ **6**

(e) $\dfrac{k}{3} + k = \dfrac{6}{3} + 6 = 2 + 6 =$ **8**

(f) $k - \dfrac{k}{6} = 6 - \dfrac{6}{6} = 6 - 1 =$ **5**

(g) $k^2 + 4 = 6 \times 6 + 4 = 36 + 4 =$ **40**
(h) $50 - k^2 = 50 - 6 \times 6 = 50 - 36 =$ **14**
(i) $k^3 - 100 = 6 \times 6 \times 6 - 100 = 216 - 100 =$ **116**
(j) $3k^2 + 20 = 3 \times 6 \times 6 + 20 = 108 + 20 =$ **128**

Exercise 3, pp. 144-145

1. (a) 3*x*   (b) 4*y*
   (c) 5*n*   (d) 6*p*
   (e) 3*x*   (f) 4*y*
   (g) 11*p*   (h) 2*e*
   (i) 4*a*   (j) 6*k*

2. (a) 2*n* + 4   (b) 5*a* + 3
   (c) 9*x* + 2   (d) 2*a* + 5
   (e) 4*d* + 2   (f) 6*f* + 9
   (g) 2*h* + 12   (h) 6*a* + 1
   (i) 2*k* + 5   (j) 5*x* + 5

Chapter 2 – Integers

Objectives

♦ Understand the meaning of positive and negative integers.
♦ Represent positive and negative integers on a horizontal or vertical number line.
♦ Compare positive and negative integers.
♦ Add positive and negative integers.
♦ Relate adding a negative integer to subtracting a positive integer of the same numerical value.

Material

♦ Counters
♦ Mental Math 14-17

Vocabulary

♦ Integers
♦ Numerical value

Notes

In *Primary Mathematics* 4 students were introduced to negative integers and learned that numbers decrease as you move to the left on the number line, even to numbers less than 0. In this chapter your student will review negative numbers and begin to use negative numbers in calculations.

Your student has probably already had informal experiences with the concept of negative numbers such as temperatures that are below 0 degrees or B.C. years in the Gregorian calendar.

Integers are natural numbers (e.g., 1, 2, 3, ...), their negatives, and 0. The value of a number without regards to its sign is its **numerical value**. It is also sometimes called its absolute value.

Students may get confused when adding and subtracting integers if they are taught to simply memorize rules without fully understanding them. It is important that your student think about real life situations in which negative numbers are used and situations in which the net result is negative. He also needs sufficient experience using a number line to add or subtract positive and negative integers before being introduced to rules. Provide as much practice with adding small positive and negative integers as needed, particularly adding positive and negative integers together, so that any rule given in the textbook is an obvious conclusion to experience, rather than a confusing bunch of words and unfamiliar terms.

At this level your student will just be adding positive and negative integers, or subtracting a positive integer from a positive or negative integer. In *Primary Mathematics* 6, students will learn to subtract a negative integer and multiply and divide integers.

(1) Review integers

Discussion

Concept p. 149

Tell your student that whole numbers, such as 1, 2, 3, and so on, are used to count objects. Numbers, including fractions and decimals, are used to count measured amounts or parts of sets, such as 2 cm or 5.5 kg or one and a half pies. Counting, whether with whole numbers or not, starts at 0; 0 objects, 0 cm, 0 kg. Sometimes, though, we want to measure something, such as distance, relative to an absolute position, such as sea level, which is "set" at 0. To do so, we can use positive and negative numbers. All distances above sea level are considered positive distances, and all distances below sea level are considered negative distances. The number 50 tells us how far from sea level, and the sign, '+' for positive or '–' for negative, tells us the direction, up or down. Ask your student if he can think of other instances where negative numbers are used. Temperature is an obvious one. Countdown to a particular point in time, such as seconds before lift-off or New Year is another example.

Tell your student that because numbers have order, they can be represented on a number line. The order on a horizontal line is traditionally from left to right; a number to the right of another one is larger than it. Negative numbers can be shown on a horizontal number line by extending it to the left. In the middle picture on the textbook page, Tom is a distance of 10 m from 0, but in the negative direction compared to Sam.

Make sure your student reads the information in the purple box. An integer is a whole number and can be negative or positive. 0 is neither negative nor positive. Decimals and fractions can be negative as well, but except for some of the optional Mental Math we will only deal with integers.

Draw a number line and use two counters of different colors, such as red and yellow. Start the yellow one at 0 and move it to the right. Ask your student what happens as you move it to the right. The numbers get larger. Start the red one at 0 and move it to the left. Ask her what happens to the numbers. The *numerical value* after the negative sign gets larger, or "more negative." However, since the we are moving to the left, the integers themselves get smaller; −10 is smaller than 0, which is smaller than +10. The numerical value, 10, tells us how far it is from 0; the negative or positive sign tells us in what direction, and allows us to compare the numbers to each other.

Tasks 1-5, pp. 150-151

Workbook

Exercise 4, pp. 146-147 (answers p. 161)

Reinforcement

Mental Math 14

---

1. (a) −25
   (b) a decrease in temperature of 40°
   (c) −$25
2. A 7
   B −1
   C −4
   D −9
3. Check number lines.
4. (a) 3 < 5   (b) 3 > −5
   (c) −3 < 5   (d) −3 > −5
   (e) −9 < −6   (f) −12 > −14
   (g) 20 > −30   (h) 15 > −15
5. (a) 5
   (b) 10
   (c) −20, 20

(2) Add positive or negative integers

Discussion

Task 6, p. 152

> 6. (a) +7
>   (b) −7

Using a number line will help your student visualize the process for adding positive or negative integers.

Task 6(a): For a more concrete image of the process, you can tell your student to imagine that the bird on p. 149 starts at the first number, and then flies higher by the amount given in the second number. We know it is going higher, or farther away from 0, by the positive sign.

Task 6(b): You can tell your student to imagine that the submarine on p. 149 starts at the first number and then submerges lower by the amount given in the second number. We know that it started below 0, and is going lower, or farther away from 0, by the negative signs on both numbers.

If necessary, do some other examples with a number line. For positive integers use a yellow counter and put it at the first number, then move it or count on for the number being added. For negative integers use a red counter and put it at the first number, then move it or count back for the negative number being added.

With enough examples, your student can conclude for herself that to add positive integers together or negative integers together, she can just add the numerical values and then put the appropriate sign in front of the numerical value of the answer.

Tell your student that use of the positive symbol is optional. A number without a symbol is positive.

Practice

Tasks 7-9, p. 152

Workbook

Exercise 5, p. 148 (answers p. 161)

> 7. (a) −8   (b) −13   (c) −93
>
> 8. −11
>
> 9. (a) −15   (b) −25   (c) −22

(3) Add positive and negative integers

Discussion

Tasks 10-13, pp. 153-154

Task 10: Although this page has a rule at the bottom of the page, your student should understand the process well enough that she can visualize it, rather than trying to remember a rule. Do as many problems as needed with a number line. You can move a counter along a drawn number line. You might want to start at 0, rather than the first number, and first move the appropriate number of spaces in the appropriate direction from 0 for the first number. When the two numbers have opposite signs, you will first move in one direction, and then in the other. Your student should observe that you move farther depending on which number has a larger numerical value, so the answer will have the same sign as that number. When the two numbers being added have opposite signs, whether the answer is positive or negative depends on how far each one is from 0, i.e., which one has the higher numerical value. Subtracting the numerical values of the two numbers tells us how much farther one is from 0 than the other.

With enough examples your student can conclude for herself that in order to add a positive integer and a negative integer, she can simply subtract the numerical values and determine whether the answer is positive or negative depending on which number has a larger numerical value.

You can also illustrate the concept with the idea of having and owing money, with the money you have being positive and the money you owe being negative. If you have \$7 and owe \$4, your net worth is +\$3, meaning that you have \$3. If you have \$4 and owe \$7, your net worth is −\$3, meaning that that you owe \$3.

Task 11: Adding a negative number is the same as subtracting the positive number with the same numerical value. If we subtract a number larger than the first number, the answer will be negative. If it is smaller, the answer will be positive. Have your student also find the answer to 10 + (−4) and 10 − 4 using the number line.

10. (a) $(+7) + (-4) = 3$
    $(-4) + (+7) = 3$
    (b) $(-6) + 2 = -4$
    $2 + (-6) = -4$

11. (a) $4 - 10 = -6$
    (b) $4 + (-10) = -6$

12. (a) 12   (b) −45   (c) −12

13. (a) \$7000 − \$3000 − \$6000
    (b) −\$2000 (\$2000 loss)

Practice

Tasks 14-15, p. 154

14. (a) −14   (b) −14   (c) −14
15. (a) −16   (b) −11   (c) −35

Workbook

Exercise 6, pp. 149-150 (answers p. 161)

Reinforcement

Mental Math 15-16

(4) Practice

Practice

   Practice B, p. 155

Reinforcement

   Extra Practice, Unit 13, Exercise 2, pp. 275-278

   Mental Math 17

Test

   Tests, Unit 13, 2A and 2B, pp. 193-196

---

1. (a) +10   (b) −82   (c) +15

2. (a) −1   (b) 6   (c) 2   (d) 0

3. (a) −7   (b) 3   (c) −1
      (d) −1   (e) −5   (f) −2

4. (a) 9   (b) 8   (c) 7
      (d) −29   (e) 11   (f) −73
      (g) −14   (h) −27   (i) −41
      (j) −15   (k) −99   (l) 260
      (m) −9   (n) −15   (o) 25
      (p) 13   (q) −35   (r) 67

5. (a) −140   (b) −162   (c) 220
      (d) −45   (e) 0   (f) −78

6. (+700) m + (−150) m + (+630 m)
   = **1180 m**

Workbook

Exercise 4, p. 146-147

1. (a) −20 m
   (b) West
   (c) −10 degrees
   (d) −20 seconds
   (e) −$20
   (f) The transaction is negative relative to the amount in savings since she is decreasing her savings. (It could also be positive with respect to the amount of cash she has.)

2. −9, −8, −7, −6, −5, −4, −3, −2, −1, 0, 1, 2, 3, 4, 5

3. (a) 8   (b) −10
   (c) −1   (d) −81

4. (a) 1   (b) −3
   (c) −29   (d) −98

5. (a) $0 < 1$ (b) $7 > -8$ (c) $-4 < -3$
   (d) $40 > -30$ (e) $-463 > -643$ (f) $458 > -458$

6. (a) 3, 1, −2, −4,
   (b) 51, −46, −50, −60

7. (a) 1, 0, −1
   (b) 0, −10, −20
   (c) −3, 0, 3

8. (a) 4   (b) 7   (c) 100

Exercise 5, p. 148

1. (a) −6   (b) −6
   (c) −8   (d) −10
   (e) 5   (f) −9

2. (a) −15   (b) 20
   (c) −108   (d) −58
   (e) −100   (f) −241
   (g) −19   (h) −30

3. (a) $n + (-4)$ (b) $(-27) + n$
      $= -5 + (-4)$ $= (-27) + (-5)$
      $= -9$ $= -32$
   (c) $n + n$ (d) $n + n + n + (-7)$
      $= (-5) + (-5)$ $= (-5) + (-5) + (-5) + (-7)$
      $= -10$ $= -22$

4. $(-\$171{,}000) + (-\$15{,}000) = -\$186{,}000$

5. $(-5°) + (-5°) + (-8°) = -18°$

Exercise 6, pp. 149-150

1. (a) 4   (b) −1
   (c) 2   (d) −4
   (e) 3   (f) −2

2. (a) −5
   (b) −13
   (c) 12
   (d) −2
   (e) −54
   (f) −29

3. (a) $n + 4$ (b) $37 + n$
      $= (-25) + 4$ $= 37 + (-25)$
      $= -21$ $= 12$
   (c) $n + 31$ (d) $n + n + 100$
      $= (-25) + 31$ $= (-25) + (-25) + 100$
      $= 6$ $= 50$

4. $1200 + (-150) + (-650) + 1050 = 1450$

5. (a) The restaurant obviously made a profit.
   (b) $\$2500 + \$1600 + (-\$700) + (-\$200) = \$3200$

Chapter 3 – Coordinate Graphs

Objectives

- ♦ Identify and graph ordered pairs in the four quadrants of the coordinate plane.
- ♦ Complete a table of values and graph the points as a straight line on a coordinate graph.
- ♦ Graph linear equations.
- ♦ Graph vertical and horizontal lines.
- ♦ Interpret information from the graph of a linear equation.

Material

- ♦ Graph paper

Vocabulary

- ♦ Coordinate graph
- ♦ Coordinates
- ♦ Ordered pair
- ♦ x-axis
- ♦ y-axis
- ♦ x-coordinate
- ♦ y-coordinate
- ♦ Quadrant
- ♦ Origin

Notes

In *Primary Mathematics* 4 students learned to plot positive points on a plane by using a coordinate graph. In Chapter 2 of this unit your student learned how to plot points for both positive and negative integers on horizontal and vertical number lines. In this chapter plotting points on a plane using coordinate graphs is reviewed and extended to negative points. Your student will also learn to graph simple linear equations.

A **coordinate graph** consists of a horizontal and a vertical number line that intersect at right angles to form a plane, a flat 2-dimensional surface. The horizontal line is called the **x-axis** and the vertical line is called the **y-axis**. The coordinate graph is divided into 4 **quadrants**. In the first quadrant, both coordinates are positive (+,+). Moving counter-clockwise from the first quadrant are the second quadrant (−,+), the third quadrant (−,−), and the fourth quadrant (+,−) respectively.

We can find any point on the graph by naming the **coordinates** of the point, which is an **ordered pair** of numbers. The first number in the pair is the **x-coordinate** and it indicates the location of the point as a distance along the x-axis. The second number in the pair is the **y-coordinate** and it indicates the location of the point as a distance along the y-axis. A common error is to mix up the coordinates.

The coordinate graph can also be used to graph equations. Your student should be somewhat familiar with the graph of a straight line in the first quadrant from line graphs and that prior knowledge can be built upon here. Equations like $y = x + 2$ and $y = x - 3$ are called linear equations because their graphs are straight lines. Your student will graph linear equations that have positive and negative slopes as well as horizontal and vertical lines. It is not important, however, that he understand slope formally at this level so terms like "positive slope," "negative slope," "slope of 0," and "undefined slope" are not used yet.

When graphing linear equations your student will need to think about the functional relationship between x and y by looking for patterns in tables of values and thinking about questions such as, "What must we do to x to get y?"

(1) Review coordinate graphs

Discussion

Concept p. 156

Depending on your student's experience, you may want to use graph paper, draw a coordinate graph, write some ordered pairs, using both positive and negative integers, and have her plot them correctly before looking at this page. To plot an ordered pair, she can start at the origin, move the distance horizontally indicated by the first number in the ordered pair, and then move the distance vertically indicated by the second number. Later, she can just start on the x-axis at the position given by the first number and move vertically up or down according to the second number of the ordered pair.

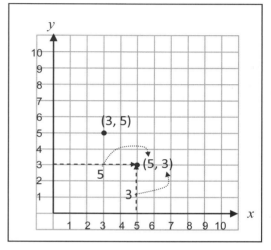

If you think your student will find plotting points in the different quadrants challenging, start with the first quadrant only, having him plot a few points in that quadrant and give you the location of some other points you plot. Then extend the x-axis to the left and ask him to give the location of a point you plot in the second quadrant. Write down some points where the first coordinate is negative and the second is positive for him to plot. Then extend the y-axis down and continue in like manner with naming some points and plotting some points in the third and fourth quadrant.

A common mistake is to mix up the order of the coordinates in the ordered pair. Help your student come up with some way to remember that the first number in an ordered pair is the distance on the x-axis.

After some practical experience plotting points, continue with the questions on p. 156, naming the 4 quadrants. The first quadrant is the upper right one, where both coordinates are positive, and the rest are numbered in a counter-clockwise direction. It is not necessary to have your student spend time memorizing which quadrant is what number; instead it is important that she be able to plot the point correctly with respect to whether each coordinate in the ordered pair is positive or negative.

Practice

Tasks 1-3, p. 157

Workbook

Exercise 7, p. 151 (answers p. 168)

1. (a) (9, 4)   (b) (0, 6)   (c) (−9, 6)   (d) (−4, 0)
   (e) (−5, −7)   (f) (0, −3)   (g) (8, −9)   (h) (10, 0)

2. (0, 0)

3. (a) 1st   (b) 3rd   (c) 4th
   (d) 2nd   (e) 2nd   (f) 3rd

(2) Graph linear equations

Activity

Refer back to the table on p. 140 of the textbook. Point out that in the expression $n + 2$ we are expressing Limei's age in terms of Angela's age. We can use any letter to stand for Angela's age. So we could also write the expression $x + 2$. When we substitute different values for x, the value of the expression $x + 2$ changes. We can assign a letter for the entire expression, such as y, and write $x + 2 = y$ or $y = x + 2$. The value of y depends on the value of x, and is always 2 more than the value of x. Point out that we have a table of values, similar to a table of data when we make a line graph. We can show the relationship between x and y visually, or graphically, using a coordinate graph. So we can plot the values for x and y that satisfy the equation $x + 2 = y$, with each pair of x and y as coordinates (x, y). Write each pair of values as ordered pairs and have your student plot the points. Tell him that we are plotting the values for x and y that satisfy the equation $x + 2 = y$.

Ask your student to connect the points in a straight line. Extend the line a bit, mark a point on the line at, for example, (3, 5) and ask him to give the coordinates for that point. If we say the first coordinate is x and the second is y, does this new pair satisfy the equation $x + 2 = y$? It does. Your student should see that any point along the line satisfies the equation. In this particular example of ages, we are assuming whole numbers, unless the birthdays fall on the same day of the year. For the general equation, $x + 2 = y$, for any point on the line, the y-coordinate is 2 more than the x-coordinate even if x is a fraction.

Repeat with the information in the table on p. 142. Since the number of apples depends on the number of packets, we assign y to the number of apples and x to the number of packets. Any point on that line, if it is drawn carefully, should satisfy the equation $y = 4x$. Again, with this example, the number of apples or bags is a whole number, but for every point on the line y is always 4 times x.

x	y	(x, y)
6	6 + 2 = 8	(6, 8)
7	7 + 2 = 9	(7, 9)
8	8 + 2 = 10	(8, 10)
9	9 + 2 = 11	(9, 11)
10	10 + 2 = 12	(10,12)
x	$y = x + 2$	

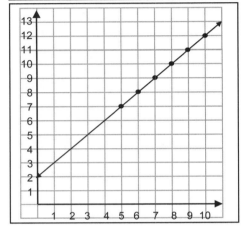

x	y	(x, y)
1	4 x 1 = 4	(1, 4)
2	4 x 2 = 8	(2, 8)
3	4 x 3 = 12	(3, 12)
4	4 x 4 = 16	(4, 16)
5	4 x 5 = 20	(5,20)
x	$y = 4x$	

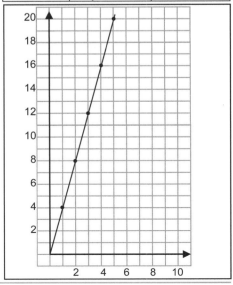

Discussion

Tasks 4-5, pp. 158-160

Task 4: This is similar to the graph made earlier, but extends the numbers to include negative integers.

Task 5: To draw the graph of a given equation, we can first find y for different values of x.

4. (a) $(-6, -4)$, $(-4, -2)$, $(-2, 0)$, $(0, 2)$, $(2, 4)$, $(4, 6)$
   (b) 2
   (c) 4
   (d) −4
   (e) 2
   (f) $y = x + 2$
   (g) Answers can vary.

5. (a)

x	−3	−2	−1	0	1	2	3
y	−6	−5	−4	−3	−2	−1	0
(x, y)	$(-3, -6)$	$(-2, -5)$	$(-1, -4)$	$(0, -3)$	$(1, -2)$	$(2, -1)$	$(3, 0)$

   (b) C

Activity

Set up a table, similar to that that at the right, and write the equation $y = -2$. Ask your student to fill in the table. She should realize that it does not matter what x is, y is always −2. Then ask her to graph the line by plotting some of the points.

Similarly, ask your student to graph the equation $x = 5$. In this case it does not matter what y is.

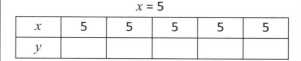

$y = -2$

x	1	2	3	4	5
y					

$x = 5$

x	5	5	5	5	5
y					

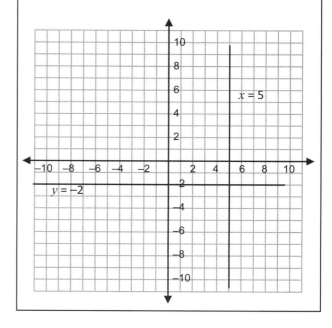

Discussion

Tasks 6-8, pp. 160-161

Be sure your student realizes that for all the points on a horizontal line, the y-coordinate is the same. Similarly, for all the points on a vertical line, the x-coordinates are the same.

The linear equation y = any constant is a horizontal line, even though the y-axis is vertical. The linear equation x = any constant is a vertical line, even thought the x-axis is horizontal. This can be confusing for some students.

6. (a) A: $(-6, 3)$
        B: $(-3, 3)$
        C: $(0, 3)$
        D: $(3, 3)$
        E: $(6, 3)$
   (b) 3
   (c) All of them.
   (d) 3
   (d) 3

7. (a) P: $(-4, 9)$
        Q: $(-4, 4)$
        R: $(-4, 0)$
        S: $(-4, -4)$
        T: $(-4, -8)$
   (b) -4
   (c) All of them.
   (d) -4
   (e) -4

Practice

Task 8, p. 161

Workbook

Exercise 8, p. 152 (answers p. 168)

8. (a) $(4, 4)$
   (b) $(4, -5)$
   (c) $(-5, -5)$
   (d) d
   (e) y
   (f) $y = x$

Practice

Tasks 9-10, pp. 162-163

Task 9(e): If your student thinks it is silly to use the graph rather than the equation to answer this problem, since it is easy to mentally divide 150 by 25, it is. The purpose of the task is just to show that you can use graphs to predict values not in the table. Some graphs are more complex than this one and not straight lines. However, calculators and computers can derive equations for lines, even those that are not straight lines, and will use the equations to find other unknown points. The answer is usually more accurate using an equation since you don't have to estimate when the value lies between two lines on the graph paper.

9. (a)

Time (min)	1	2	3	4	5	x
Amount of water (gal)	25	50	75	100	125	$25x$

(b) $y = 25x$
(c) (1, 25), (2, 50), (3, **75**), (4, **100**), (5, **125**)
(d) yes
(e) 6 min

10. (a)

x	0	1	2	3	4	5
y	−3	−1	1	3	5	7
(x, y)	(0, −3)	(1, −1)	(2, 1)	(3, 3)	(4, 5)	(5, 7)

(b) b
(c) −9
(d) −1

Task 10(c): Your student might also think this one is easier to answer using the equation, even though she has not formally learned to multiply a positive number by a negative number. Sometimes using a graph to solve an equation is easier if a graph is already available and an estimated answer is sufficient if the equation is complex.

Workbook

Exercise 9, pp. 153-154 (answers p. 168)

Reinforcement

Extra Practice, Unit 13, Exercise 3, pp. 279-282

Test

Tests, Unit 13, 3A and 3B, pp. 197-204

Workbook

Exercise 7, p. 151

1.

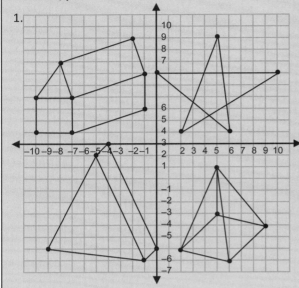

Exercise 8, p. 152

1. (a)

x	0	1	2	3	4	5	6
y	4	3	2	1	0	−1	−2
(x, y)	(0, 4)	(1, 3)	(2, 2)	(3, 1)	(4, 0)	(5, −1)	(6, −2)

(b)

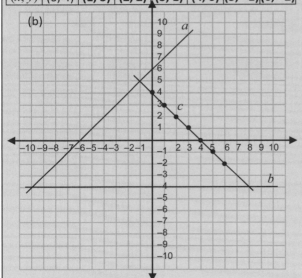

(c) (−1, 5), (−10, −4), (8, −4)

(d) $\frac{1}{2}$ x 18 units x 9 units = **81 square units**

(e) −4

(f) line a

Exercise 9, pp. 153-154

1. (a)

1	2	3	4	5	x
15	30	45	60	75	15x

(b) $y = 15x$

(c) (1, 15), (2, 30), (3, 45), (4, 60), (5, 75)

(d)

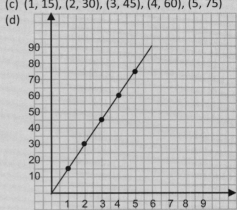

(e) 6 gal

(f) $x = \frac{y}{15} = \frac{120}{15} = 8$ **8 gal**

2. (a)

0	1	2	3	4	5
2	3.5	5	6.5	8	9.5
(0, 2)	(1, 3.5)	(2, 5)	(3, 6.5)	(4, 8)	(5, 9.5)

(b)

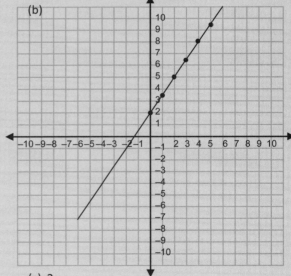

(c) 2

(d) −4

(e) $-1\frac{1}{2}$

Review 13

Review

Review 13, pp. 164-169

Problem 13(b) should be changed to 95 – ___ > 18 if it is not already changed in your printing. There is no greatest whole number for 95 + ___ > 18.

Workbook

Review 13, pp. 155-160 (answers p. 171)

Tests

Tests, Units 1-13, Cumulative Tests A and B, pp. 205-212

1. (a) 10,000
   (b) 0.06

2. 4

3. (a) 600.83
   (b) 8.04
   (c) 7.031
   (d) 45.009

4. (a) 0.57   (b) 0.83   (c) 3.22

5. 24

6. (a) 4.998
   (b) 7.65

7. (a) 40   (b) 52   (c) 450
   (d) 7   (e) −7.5   (f) 13

8. (a) $9y + 7$ (b) $7y + 7$

9. (a) $(12 + m + 5)$ years = **$(17 + m)$ years**
   (b) $(17 + m)$ years = $(17 + 20)$ years = **37 years**

10. (a) $(20 + 2x)$ cm
    (b) $(20 + 2 \times 6)$ cm = **32 cm**

11. Side of each square: 7 cm ($7 \times 7 = 49$)
    Perimeter = 10×7 cm = **70 cm**

12. (a) $64 - \underline{(24 - 18)} \times 10$ (b) $\underline{15 \div 3} + \underline{(9 - 6)} \times 4$
        $= 64 - \underline{6 \times 10}$ $= 5 + \underline{3 \times 4}$
        $= 64 - 60$ $= 5 + 12$
        $= \mathbf{4}$ $= \mathbf{17}$

13. (a) 4   (b) 76 (See note at left.)

14. (a) 1   (b) 10
    (c) 3   ($3\frac{3}{5} = 3\frac{9}{15}$; $4\frac{4}{15} = 3\frac{19}{15}$; so we need to add $\frac{9}{15}$, or $\frac{2}{3}$)
    (d) 15.7   (? parts x 0.2 in each part = 3.14 total; total ÷ 0.2 = ?)

15. $90 \rightarrow 5$ min
    $270 \rightarrow 5$ min x 3 = **15 min**
    It will take 15 min to print 270 copies.

16. $160 \rightarrow 2$ min
    $400 \rightarrow \frac{2}{160}$ min x 400 = **5 min**
    It will take 5 min to cap 400 bottles.

17. 90 m : 60 m = **3 : 2**

(continued next page)

18. $(2 \times \$4.90) + (13 \times \$2.50) - \$35$
   $= \$9.80 + \$32.5 - \$35$
   $= \mathbf{\$7.30}$
   The amount of money made is $7.30.

19. $\dfrac{98}{200} \times 100\% = \mathbf{49\%}$
   49% of the coins are commemorative coins.

20. (a) $\dfrac{20}{8 \times 50} \times 100\% = \mathbf{5\%}$
   5% of the pomegranates were rotten.
   (b) Number of pomegranates sold:
   $(8 \times 50) - 20 = 380$
   Money on sales: $380 \times \$1.50 = \$570$
   Money made: $\$570 - \$400 = \mathbf{\$170}$

21. $\dfrac{1}{4} + \dfrac{1}{2} \times \dfrac{3}{4} = \dfrac{2}{8} + \dfrac{3}{8} = \dfrac{5}{8}$
   He spent $\dfrac{5}{8}$ of his money.

22. 8 gal $\rightarrow$ 1 min
   200 gal $\rightarrow \dfrac{1}{8}$ min x 200 = **25 min**
   It will take 25 min to fill the tank.

23. Total for 3 numbers: $45 \times 3 = 135$
   Total for 2 numbers: $27 \times 2 = 54$
   Value of 3rd number: $135 - 54 = \mathbf{81}$

24. $\dfrac{3}{4}$ ℓ $\rightarrow$ 4 glasses
   3 $\rightarrow$ 4 glasses x 4 = **16** glasses
   3 liters will fill 16 glasses.

25. $2 $\rightarrow$ 5 oranges
   $24 $\rightarrow$ 5 x 12 = **60** oranges
   $24 will buy 60 oranges.

26. (a) $\angle x = 90° - 37° = \mathbf{53°}$
   (b) $\angle x = 360° - 90° - 105° - 47° = \mathbf{118°}$

27. $100\% - 25\% - 20\% - 15\% = \mathbf{40\%}$
   40% walk to school.

28. (a) $\dfrac{1}{4}$ of the people were boys.
   (b) $\dfrac{1}{2} - \dfrac{1}{8} - \dfrac{1}{4} = \dfrac{1}{8}$
   $\dfrac{1}{8}$ of the people were women.
   (c) $120 \times 8 = \mathbf{960}$
   There were 960 people on the cruise.

29. (a) $\$3000 - \$1200 = \mathbf{\$1800}$
   Sales increased $1800 between Feb. and Mar.
   (b) $\$2000 + \$1200 + \$3000 + \$2400 + \$800 = \$9400$
   $\$9400 \div 5 = \mathbf{\$1880}$
   The average sales was $1880 per month.
   (c) $\$2400 \div \$4 = \mathbf{600}$
   600 T-shirts were sold in April.

30. (a) 1st   (b) 3rd   (c) 4th
   (d) 2nd   (e) 2nd   (f) 3rd

31. (a) $(-4, 0)$
   (b) 5
   (c) Table can vary.
   Example at right.
   $y = x + 4$

x	0	−4	4
y	4	0	8

32. (a) $2 + 8 + 9 + 5 = \mathbf{24}$
   (b) 10
   (c) 5
   (d) 80-89
   (e) 60-69
   (f) $\dfrac{8}{24} \times 100\% = \mathbf{33\dfrac{1}{3}\%}$
   (g) $\dfrac{9}{24} = \dfrac{3}{8}$

Workbook

Review 13, pp. 155-160

1. 258.8

2. 0.5 yd is less than 3 ft, and 36 in. is 3 ft, so **37 in.** is greatest in value.

3. (a) 3 qt 3 c
   (b) 22 qt 1 pt
   (c) 13 gal 0 qt
   (d) 3 ft 2 in.

4. −60

5. (a) −5   (b) $\frac{2}{3}$ (c) 171

6. (a) $2a$ (b) $3a$ (c) $16a + 4$

7. (a) −4   (b) 0   (c) −9
   (d) −20   (e) −65   (f) −8

8. 2 + 3 + 5 + 7 + 11 + 13 + 17 + 19 + 23 + 29 = **129**

9. (a) $\frac{1}{2}$ x 15 cm x 8 cm   (b) $\frac{1}{2}$ x 8 ft x 12 ft
   $\quad$ = **60 cm²** $\qquad\qquad$ = **48 ft²**

10. $\frac{2}{5}$ → 14 gal $\qquad$ Or: 14 gal ÷ $\frac{2}{5}$ = 35 gal

    $\frac{1}{5}$ → $\frac{14}{2}$ gal = 7 gal

    $\frac{5}{5}$ → 7 x 5 gal = **35 gal**

11. (a) 11.3   (b) 249.2   (c) 160.5

12. (a) $1\frac{3}{5} + 4\frac{3}{8} = 5\frac{24}{40} + \frac{15}{40} = 5\frac{39}{40}$

    (b) $\frac{\cancel{4}^{\,1}}{\cancel{8}_{\,1}} \times \frac{\cancel{5}^{\,1}}{\cancel{8}_{\,2}} = \frac{1}{2}$

    (c) $12 ÷ \frac{3}{5} = \cancel{12}^{\,4} \times \frac{5}{\cancel{3}_{\,1}} = 20$

13. Total cost of 3 mugs: $4 x 3 = $12
    Cost of third mug: $12 − $p − $3 = **$(9 − p)**

14. 5 units total
    2 units = 9 cm
    1 unit = 4.5 cm
    3 units = 3 x 4.5 cm = 13.5 cm
    Area = 9 cm x 13.5 cm = **121.5 cm²**

15.

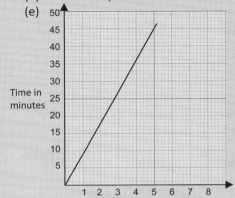

Divide fifths for boys by 2 and halves for girls by 5. 20 units total. 4 units of boys and 5 units of girls take the bus.

$\frac{9}{20}$ of all the students take the bus.

16. ∠CBD = 116° $\qquad$ (vert. opp. ∠'s)
    ∠BCD = (180° − 116°) ÷ 2 $\quad$ (iso. Δ, ∠ sum of Δ)
    $\qquad$ = **32°**

17. (a) 25%
    (b) 25% − 10% = **15%**
    (c) $200 x 2 = **$400**

18. As in the lessons, assume that the line that looks straight is a diameter.
    (a) 6A = 6C; So 6C donated **25%**.
    (b) $55 x 4 = **$220**
    (c) (2 x $55) − $44 = **$66**
    (d) $55 : $44 = **5 : 4**

19. (a) 125 ℓ
    (b) 10 min

20. (a) 5 pages → 45 min
    $\qquad$ 1 page → $\frac{45}{5}$ min = **9 min**
    (b) 9 min → 1 page
    $\qquad$ 1 min → $\frac{1}{9}$ page
    (c) $y = 9x$
    (d) 9 x 20 = 180; **180 min**
    (e)

    (f) 3 pages
    (g) 36 min

Mental Math 1	Mental Math 2	Mental Math 3	Mental Math 4	Mental Math 5	Mental Math 6
4.9+0.05=**4.95**	0.099+**0.901**=1	3÷4=**0.75**	0.4 x 10 = **4**	72 ÷ 90 = **0.8**	31 ÷ 100 = 3.1 ÷ **10**
4.95+0.05=**5**	0.099+**0.001**=0.1	36÷48=**0.75**	0.7 x 20 = **14**	12 ÷ 600 = **0.02**	3.1 ÷ 10 = 310 ÷ **1000**
4.982+0.05=**5.032**	0.756+**0.244**=1	1÷6≈**0.167**	0.02 x 100 = **2**	63 ÷ 9000 = **0.007**	0.31 x 100 = 0.031 x **1000**
3.42−0.08=**3.34**	0.042+**0.958**=1	15÷90≈**0.167**	0.06 x 500 = **30**	5.6 ÷ 80 = **0.07**	0.31 x 1000 = 31 x **10**
3.458−0.08=**3.378**	5.52+**0.48**=6	1÷8=**0.125**	0.03 x 1000 = **30**	0.3 ÷ 50 = **0.006**	310 ÷ 100 = 0.31 x **10**
3.4−0.08=**3.32**	3.651+**0.349**=4	3÷8=**0.375**	0.07 x 9000 = **630**	420 ÷ 700 = **0.6**	3100 ÷ 10 = 0.31 x **1000**
3−0.08=**2.92**	0.033+**0.067**=0.1	33÷88=**0.375**	0.003 x 9000 = **27**	81 ÷ 90 = **0.9**	0.31 x 100 = 3100 ÷ **100**
9.01+0.009=**9.019**	0.007+**0.003**=0.01	1÷3≈**0.333**	60 x 0.06 = **3.6**	480 ÷ 8000 = **0.06**	0.04 x 100 = **0.004** x 1000
0.141−0.06=**0.081**	0.742+**0.258**=1	4÷3≈**1.333**	0.09 x 800 = **72**	240 ÷ 600 = **0.4**	0.4 x 10 = **0.04** x 100
8.842+0.6=**9.442**	4.12+**0.88**=5	95÷15≈**6.333**	0.7 x 80 = **56**	4.5 ÷ 900 = **0.005**	4 ÷ 100 = **40** ÷ 1000
0.142−0.006=**0.136**	8.552+**0.448**=9	3÷5=**0.6**	0.06 x 400 = **240**	36 ÷ 6000 = **0.006**	0.04 ÷ 10 = **4** ÷ 1000
7.549−0.003=**7.546**	0.091+**0.009**=0.1	8÷5=**1.6**	4000 x 0.05 = **200**	280 ÷ 400 = **0.7**	400 ÷ 1000 = **0.04** x 10
3.274+0.09=**3.364**	0.141+**0.859**=1	48÷30=**1.6**	600 x 0.9 = **540**	0.09 ÷ 30 = **0.003**	48 ÷ 800 = 4.8 ÷ 80
1.47+0.8=**2.27**	0.205+**0.795**=1	2÷3≈**0.667**	0.008 x 20 = **0.16**	2.4 ÷ 200 = **0.012**	480 ÷ 800 = 48 ÷ 80
9.105−0.5=**8.605**	0.027+**0.973**=1	32÷48=**0.667**	0.03 x 600 = **18**	0.66 ÷ 60 = **0.011**	560 ÷ 8000 = 0.07
2.355+0.005=**2.36**	7.71+**1.29**=9	2÷5=**0.4**	0.5 x 3000 = **1500**	10 ÷ 500 = **0.02**	0.6 x 700 = 420
1.5+0.09=**1.59**	6.62+**2.38**=9	22÷5=**4.4**	0.08 x 500 = **40**	3.6 ÷ 300 = **0.012**	280 ÷ 400 = 0.7
1.5−0.09=**1.41**	0.023+**1.977**=2	57÷5=**11.4**	0.06 x 70 = **4.2**	48 ÷ 40 = **1.2**	0.03 x 50 = 1.5
7.482+0.99=**8.472**	1.19+**4.81**=6	5÷2=**2.5**	90 x 0.09 = **8.1**	200 ÷ 4000 = **0.05**	30 ÷ 500 = 0.06
7.482−0.99=**6.492**	0.035+**2.965**=3	125÷50=**2.5**	0.2 x 5000 = **1000**	85 ÷ 500 = **0.17**	0.007 x 200 = 1.4

Mental Math 7	Mental Math 8	Mental Math 9	Mental Math 10	Mental Math 11	Mental Math 12
9 x 0.3 = **2.7**	15 ÷ 0.5 = **30**	0.35 ℓ = **350** ml	$\frac{7}{20}$ = **35%**	$\frac{86}{200}$ = **43%**	10% of 500 = **50**
0.7 x 0.09 = **0.063**	6.4 ÷ 0.8 = **8**	0.25 ft = **3** in.	$\frac{9}{25}$ = **36%**	$\frac{123}{300}$ = **41%**	1% of 500 = **5**
0.002 x 0.5 = **0.001**	0.42 ÷ 0.7 = **0.6**	0.5 min = **30** s	$\frac{4}{5}$ = **80%**	$\frac{7}{700}$ = **1%**	12% of 500 = **60**
8 x 0.06 = **0.48**	0.03 ÷ 0.6 = **0.05**	4 oz = **0.25** lb	$\frac{21}{25}$ = **84%**	$\frac{248}{400}$ = **62%**	62% of 500 = **310**
0.03 x 0.8 = **0.024**	5.4 ÷ 0.09 = **60**	0.5 yr = **6** mnths	$\frac{3}{4}$ = **75%**	$\frac{65}{500}$ = **13%**	65% of 500 = **325**
0.7 x 0.7 = **0.49**	0.24 ÷ 0.04 = **6**	1 qt = **0.25** gal	$\frac{7}{10}$ = **70%**	$\frac{36}{600}$ = **6%**	15% of 600 = **90**
2 x 0.007 = **0.014**	16 ÷ 0.002 = **8000**	9 cm = **0.09** m	$\frac{89}{100}$ = **89%**	$\frac{50}{250}$ = **20%**	6% of 600 = **36**
0.08 x 0.05 = **0.004**	2 ÷ 0.005 = **400**	8.45 kg = **8450** g	$\frac{3}{5}$ = **60%**	$\frac{819}{900}$ = **91%**	16% of 600 = **96**
0.5 x 0.5 = **0.25**	8.1 ÷ 0.009 = **900**	1.25 ft = **15** in.	$\frac{19}{20}$ = **95%**	$\frac{168}{800}$ = **21%**	42% of 200 = **84**
0.9 x 0.04 = **0.036**	0.049 ÷ 0.7 = **0.07**	21 cm = **0.21** m	$\frac{11}{20}$ = **55%**	$\frac{54}{300}$ = **18%**	35% of 300 = **105**
40 x 0.007 = **0.28**	0.004 ÷ 0.8 = **0.005**	9 mnths = **0.75** yr	$\frac{4}{16}$ = **25%**	$\frac{156}{200}$ = **78%**	40% of 80 = **32**
0.5 x 0.4 = **0.2**	270 ÷ 0.009 = **30,000**	7.04 m = **704** cm	$\frac{15}{30}$ = **50%**	$\frac{132}{400}$ = **33%**	25% of 360 = **90**
0.03 x 0.2 = **0.006**	0.025 ÷ 0.5 = **0.05**	48 ml = **0.048** ℓ	$\frac{12}{40}$ = **30%**	$\frac{415}{500}$ = **83%**	15% of 80 = **12**
0.007 x 5 = **0.035**	4.5 ÷ 0.5 = **9**	450 cm = **4.5** m	$\frac{42}{60}$ = **70%**	$\frac{444}{600}$ = **74%**	15% of 420 = **63**
0.8 x 0.07 = **0.056**	21 ÷ 0.003 = **7000**	590 g = **0.59** kg	$\frac{37}{50}$ = **74%**	$\frac{238}{700}$ = **34%**	8% of 800 = **64**
40 x 0.008 = **0.32**	3.2 ÷ 0.04 = **80**	45 min = **0.75** hr	$\frac{23}{25}$ = **92%**	$\frac{792}{800}$ = **99%**	8% of 1000 = **80**
0.09 x 0.3 = **0.027**	0.35 ÷ 0.005 = **70**	18 in. = **1.5** ft	$\frac{27}{50}$ = **54%**	$\frac{307}{614}$ = **50%**	73% of 1000 = **730**
0.004 x 4 = **0.016**	0.024 ÷ 0.8 = **0.03**	0.02 m = **2** cm			20% of 380 = **76**
0.2 x 0.02 = **0.004**	0.048 ÷ 0.06 = **0.8**	0.75 qt = **3** c			60% of 60 = **36**
0.006 x 20 = **0.12**	1800 ÷ 0.006 = **300,000**				21% of 2100 = **441**

Mental Math 13	Mental Math 14	Mental Math 15		Mental Math 16	Mental Math 17
Find the value when $n = 7$.	$-1 < 1$	$(+2) + (+3) = \mathbf{5}$		Find the value when $n = -12$.	$-7 + 3 < 7 + 3$
$37 + n = \mathbf{44}$	$4 > -1$	$(-2) + (-3) = \mathbf{-5}$		$n + 8 = \mathbf{-4}$	$6 - 3 > 3 - 6$
$n - 5 = \mathbf{2}$	$0 > -10$	$(+2) + (-3) = \mathbf{-1}$		$n - 9 = \mathbf{-21}$	$-2 - 7 > -2 - 8$
$11n = \mathbf{77}$	$-4 < -3$	$(-2) + (+3) = \mathbf{1}$		$n - 18 = \mathbf{-30}$	$3 - 7 - 5 < 7 - 3 - 5$
$3n + 2n = \mathbf{35}$	$-10 > -12$	$12 + (-3) = \mathbf{9}$		$15 + n = \mathbf{3}$	$-3 - 7 + 5 < 3 - 5 + 7$
$28n - 18n = \mathbf{70}$	$16 > -18$	$1 - 10 = \mathbf{-9}$		$-25 + n = \mathbf{-37}$	$-5 + 7 - 3 = -3 - 5 + 7$
$3n + 3 = \mathbf{24}$	$-23 < -21$	$(-16) + 5 = \mathbf{-11}$		$n + 100 = \mathbf{88}$	$-13 - 19 < 34 - 6$
$8n - 7 = \mathbf{49}$	$-106 < 99$	$-40 + 12 = \mathbf{-28}$		$n + (-100) = \mathbf{-112}$	$-10 - 3 < 3 + (-10)$
$3n + 1 + 4n = \mathbf{50}$	$-123 > -132$	$-7 - 2 = \mathbf{-9}$		$-100 + n = \mathbf{-112}$	$-11 - 13 = 11 - 35$
$4n - 7 - n = \mathbf{14}$	$-62 > -162$	$8 + (-17) = \mathbf{-9}$		$100 + n = \mathbf{88}$	$-66 - 98 < -98 - 53$
$\frac{9n}{3} = \mathbf{21}$	$272 < 727$	$(-7) + 14 = \mathbf{7}$		$n + 10 - 6 = \mathbf{-8}$	$71 + 38 > -71 + 38$
$\frac{20n}{5} + 12 = \mathbf{40}$	$-272 > -727$	$(+56) + (-20) = \mathbf{36}$		$(-7) + n + 19 = \mathbf{0}$	$10 - 13 = 55 - 58$
$\frac{8n}{2} - n = \mathbf{21}$	$-999 > -1000$	$-30 + 42 = \mathbf{12}$		$(-7) - 2 + n = \mathbf{-21}$	$-32 - 87 < -41 - 69$
$\frac{n-1}{2} = \mathbf{3}$	$-1.23 > -2.13$	$(-21) + (-3) = \mathbf{-24}$		$n + n = \mathbf{-24}$	$-87 - 53 + 76 = -53 + 76 - 87$
$\frac{6n+7n}{7} = \mathbf{13}$	$-7.3 < -0.73$	$(-11) + (+32) = \mathbf{21}$		$n + n - 100 = \mathbf{-124}$	$39 + 82 - 47 > -47 + 39 - 82$
$\frac{2n+28}{14} = \mathbf{3}$	$-\frac{5}{8} < -\frac{3}{8}$	$100 + (-72) = \mathbf{28}$		$100 + n + n = \mathbf{76}$	$653 - 563 > 563 - 653$
$n - \frac{4n-3}{5} = \mathbf{2}$	$\frac{1}{2} > -\frac{1}{3}$	$54 - 100 = \mathbf{-46}$		$n + 100 + n + 4 = \mathbf{80}$	$653 - 635 > 536 - 563$
$10 + 10n - 20 = \mathbf{60}$	$-\frac{3}{5} < -\frac{3}{7}$	$(-99) + 16 = \mathbf{-83}$		$n + n + n - 15 = \mathbf{-51}$	$-1000 - 76 < 76 - 1000$
$n + 10 - 2n = \mathbf{3}$	$-\frac{3}{5} > -\frac{5}{8}$	$-43 + 199 = \mathbf{156}$		$n + (-40) + n + n = \mathbf{-76}$	$-1000 + 173 < 827 - 1000$
		$-72 - 399 = \mathbf{-471}$		$1000 + n + n + n = \mathbf{964}$	

Blank Page

Mental Math 1	Mental Math 2	Mental Math 3
4.9 + 0.05 = _____	0.099 + _____ = 1	3 ÷ 4 = _____
4.95 + 0.05 = _____	0.099 + _____ = 0.1	36 ÷ 48 = _____
4.982 + 0.05 = _____	0.756 + _____ = 1	1 ÷ 6 ≈ _____
3.42 − 0.08 = _____	0.042 + _____ = 1	15 ÷ 90 ≈ _____
3.458 − 0.08 = _____	5.52 + _____ = 6	1 ÷ 8 = _____
3.4 − 0.08 = _____	3.651 + _____ = 4	3 ÷ 8 = _____
3 − 0.08 = _____	0.033 + _____ = 0.1	33 ÷ 88 = _____
9.01 + 0.009 = _____	0.007 + _____ = 0.01	1 ÷ 3 ≈ _____
0.141 − 0.06 = _____	0.742 + _____ = 1	4 ÷ 3 ≈ _____
8.842 + 0.6 = _____	4.12 + _____ = 5	95 ÷ 15 ≈ _____
0.142 − 0.006 = _____	8.552 + _____ = 9	3 ÷ 5 = _____
7.549 − 0.003 = _____	0.091 + _____ = 0.1	8 ÷ 5 = _____
3.274 + 0.09 = _____	0.141 + _____ = 1	48 ÷ 30 = _____
1.47 + 0.8 = _____	0.205 + _____ = 1	2 ÷ 3 ≈ _____
9.105 − 0.5 = _____	0.027 + _____ = 1	32 ÷ 48 ≈ _____
2.355 + 0.005 = _____	7.71 + _____ = 9	2 ÷ 5 = _____
1.5 + 0.09 = _____	6.62 + _____ = 9	22 ÷ 5 = _____
1.5 − 0.09 = _____	0.023 + _____ = 2	57 ÷ 5 = _____
7.482 + 0.99 = _____	1.19 + _____ = 6	5 ÷ 2 = _____
7.482 − 0.99 = _____	0.035 + _____ = 3	125 ÷ 50 = _____

Mental Math 4	Mental Math 5	Mental Math 6
0.4 x 10 = _____	72 ÷ 90 = _____	31 ÷ 100 = 3.1 ÷ _____
0.7 x 20 = _____	12 ÷ 600 = _____	3.1 ÷ 10 = 310 ÷ _____
0.02 x 100 = _____	63 ÷ 9000 = _____	0.31 x 100 = 0.031 x _____
0.06 x 500 = _____	5.6 ÷ 80 = _____	0.31 x 1000 = 31 x _____
0.03 x 1000 = _____	0.3 ÷ 50 = _____	310 ÷ 100 = 0.31 x _____
0.07 x 9000 = _____	420 ÷ 700 = _____	3100 ÷ 10 = 0.31 x _____
0.003 x 9000 = _____	81 ÷ 90 = _____	0.31 x 100 = 3100 ÷ _____
60 x 0.06 = _____	480 ÷ 8000 = _____	0.04 x 100 = _____ x 1000
0.09 x 800 = _____	240 ÷ 600 = _____	0.4 x 10 = _____ x 100
0.7 x 80 = _____	4.5 ÷ 900 = _____	4 ÷ 100 = _____ ÷ 1000
0.06 x 400 = _____	36 ÷ 6000 = _____	0.04 ÷ 10 = _____ ÷ 1000
4000 x 0.05 = _____	280 ÷ 400 = _____	400 ÷ 1000 = _____ x 10
600 x 0.9 = _____	0.09 ÷ 30 = _____	_____ ÷ 800 = 4.8 ÷ 80
0.008 x 20 = _____	2.4 ÷ 200 = _____	_____ ÷ 800 = 48 ÷ 80
0.03 x 600 = _____	0.66 ÷ 60 = _____	_____ ÷ 8000 = 0.07
0.5 x 3000 = _____	10 ÷ 500 = _____	_____ x 700 = 420
0.08 x 500 = _____	3.6 ÷ 300 = _____	_____ ÷ 400 = 0.7
0.06 x 70 = _____	48 ÷ 40 = _____	_____ x 50 = 1.5
90 x 0.09 = _____	200 ÷ 4000 = _____	_____ ÷ 500 = 0.06
0.2 x 5000 = _____	85 ÷ 500 = _____	_____ x 200 = 1.4

Mental Math 7	Mental Math 8	Mental Math 9
9 x 0.3 = _____	15 ÷ 0.5 = _____	0.35 ℓ = _____ ml
0.7 x 0.09 = _____	6.4 ÷ 0.8 = _____	0.25 ft = _____ in.
0.002 x 0.5 = _____	0.42 ÷ 0.7 = _____	0.5 min = _____ s
8 x 0.06 = _____	0.03 ÷ 0.6 = _____	4 oz = _____ lb
0.03 x 0.8 = _____	5.4 ÷ 0.09 = _____	0.5 yr = _____ mnths
0.7 x 0.7 = _____	0.24 ÷ 0.04 = _____	1 qt = _____ gal
2 x 0.007 = _____	16 ÷ 0.002 = _____	9 cm = _____ m
0.08 x 0.05 = _____	2 ÷ 0.005 = _____	8.45 kg = _____ g
0.5 x 0.5 = _____	8.1 ÷ 0.009 = _____	1.25 ft = _____ in.
0.9 x 0.04 = _____	0.049 ÷ 0.7 = _____	21 cm = _____ m
40 x 0.007 = _____	0.004 ÷ 0.8 = _____	9 mnths = _____ yr
0.5 x 0.4 = _____	270 ÷ 0.009 = _____	7.04 m = _____ cm
0.03 x 0.2 = _____	0.025 ÷ 0.5 = _____	48 ml = _____ ℓ
0.007 x 5 = _____	4.5 ÷ 0.5 = _____	450 cm = _____ m
0.8 x 0.07 = _____	21 ÷ 0.003 = _____	590 g = _____ kg
40 x 0.008 = _____	3.2 ÷ 0.04 = _____	45 min = _____ hr
0.09 x 0.3 = _____	0.35 ÷ 0.005 = _____	1.5 hr = _____ min
0.004 x 4 = _____	0.024 ÷ 0.8 = _____	18 in. = _____ ft
0.2 x 0.02 = _____	0.048 ÷ 0.06 = _____	0.02 m = _____ cm
0.006 x 20 = _____	1800 ÷ 0.006 = _____	0.75 qt = _____ c

Mental Math

Mental Math 10	Mental Math 11	Mental Math 12
$\frac{7}{20}$ = _____ %	$\frac{86}{200}$ = _____ %	10% of 500 = _____
$\frac{9}{25}$ = _____ %	$\frac{123}{300}$ = _____ %	1% of 500 = _____
$\frac{4}{5}$ = _____ %	$\frac{7}{700}$ = _____ %	12% of 500 = _____
$\frac{21}{25}$ = _____ %	$\frac{248}{400}$ = _____ %	62% of 500 = _____
$\frac{3}{4}$ = _____ %	$\frac{65}{500}$ = _____ %	65% of 500 = _____
$\frac{7}{10}$ = _____ %	$\frac{36}{600}$ = _____ %	15% of 600 = _____
$\frac{89}{100}$ = _____ %	$\frac{50}{250}$ = _____ %	6% of 600 = _____
$\frac{3}{5}$ = _____ %	$\frac{819}{900}$ = _____ %	16% of 600 = _____
$\frac{19}{20}$ = _____ %	$\frac{168}{800}$ = _____ %	42% of 200 = _____
$\frac{11}{20}$ = _____ %	$\frac{54}{300}$ = _____ %	35% of 300 = _____
$\frac{4}{16}$ = _____ %	$\frac{156}{200}$ = _____ %	40% of 80 = _____
$\frac{15}{30}$ = _____ %	$\frac{132}{400}$ = _____ %	25% of 360 = _____
$\frac{12}{40}$ = _____ %	$\frac{415}{500}$ = _____ %	15% of 80 = _____
$\frac{42}{60}$ = _____ %	$\frac{444}{600}$ = _____ %	15% of 420 = _____
$\frac{37}{50}$ = _____ %	$\frac{238}{700}$ = _____ %	8% of 800 = _____
$\frac{23}{25}$ = _____ %	$\frac{792}{800}$ = _____ %	8% of 1000 = _____
$\frac{27}{50}$ = _____ %	$\frac{307}{614}$ = _____ %	73% of 1000 = _____
		20% of 380 = _____
		60% of 60 = _____
		21% of 2100 = _____

Mental Math 13	Mental Math 14	Mental Math 15
Find the value when $n = 7$.	Fill in ◯ with > or <.	$(+2) + (+3) = $ _____
$37 + n = $ _____	-1 ◯ 1	$(-2) + (-3) = $ _____
$n - 5 = $ _____	4 ◯ -1	$(+2) + (-3) = $ _____
$11n = $ _____	0 ◯ -10	$(-2) + (+3) = $ _____
$3n + 2n = $ _____	-4 ◯ -3	$12 + (-3) = $ _____
$28n - 18n = $ _____	-10 ◯ -12	$1 - 10 = $ _____
$3n + 3 = $ _____	16 ◯ -18	$(-16) + 5 = $ _____
$8n - 7 = $ _____	-23 ◯ -21	$-40 + 12 = $ _____
$3n + 1 + 4n = $ _____	-106 ◯ 99	$-7 - 2 = $ _____
$4n - 7 - n = $ _____	-123 ◯ -132	$8 + (-17) = $ _____
$\dfrac{9n}{3} = $ _____	-62 ◯ -162	$(-7) + 14 = $ _____
$\dfrac{20n}{5} + 12 = $ _____	272 ◯ 727	$(+56) + (-20) = $ _____
$\dfrac{8n}{2} - n = $ _____	-272 ◯ -727	$-30 + 42 = $ _____
$\dfrac{n-1}{2} = $ _____	-999 ◯ -1000	$(-21) + (-3) = $ _____
$\dfrac{6n+7n}{7} = $ _____	-1.23 ◯ -2.13	$(-11) + (+32) = $ _____
$\dfrac{2n+28}{14} = $ _____	-7.3 ◯ -0.73	$100 + (-72) = $ _____
$n - \dfrac{4n-3}{5} = $ _____	$-\dfrac{5}{8}$ ◯ $-\dfrac{3}{8}$	$54 - 100 = $ _____
$10 + 10n - 20 = $ _____	$\dfrac{1}{2}$ ◯ $-\dfrac{1}{3}$	$(-99) + 16 = $ _____
$n + 10 - 2n = $ _____	$-\dfrac{3}{5}$ ◯ $-\dfrac{3}{7}$	$-43 + 199 = $ _____
	$-\dfrac{3}{5}$ ◯ $-\dfrac{5}{8}$	$-72 - 399 = $ _____

Mental Math 16	Mental Math 17

Find the value when $n = -12$.

$n + 8 =$ _____

$n - 9 =$ _____

$n - 18 =$ _____

$15 + n =$ _____

$-25 + n =$ _____

$n + 100 =$ _____

$n + (-100) =$ _____

$-100 + n =$ _____

$100 + n =$ _____

$n + 10 - 6 =$ _____

$(-7) + n + 19 =$ _____

$(-7) - 2 + n =$ _____

$n + n =$ _____

$n + n - 100 =$ _____

$100 + n + n =$ _____

$n + 100 + n + 4 =$ _____

$n + n + n - 15 =$ _____

$n + (-40) + n + n =$ _____

$1000 + n + n + n =$ _____

Fill in $\bigcirc$ with >, <, or =.

$-7 + 3 \bigcirc 7 + 3$

$6 - 3 \bigcirc 3 - 6$

$-2 - 7 \bigcirc -2 - 8$

$3 - 7 - 5 \bigcirc 7 - 3 - 5$

$-3 - 7 + 5 \bigcirc 3 - 5 + 7$

$-5 + 7 - 3 \bigcirc -3 - 5 + 7$

$-13 - 19 \bigcirc 34 - 6$

$-10 - 3 \bigcirc 3 + (-10)$

$-11 - 13 \bigcirc 11 - 36$

$-66 - 98 \bigcirc -98 - 53$

$71 + 38 \bigcirc -71 + 38$

$10 - 13 \bigcirc 55 - 58$

$-32 - 87 \bigcirc -41 - 69$

$-87 - 53 + 76 \bigcirc -53 + 76 - 87$

$39 + 82 - 47 \bigcirc -47 + 39 - 82$

$653 - 563 \bigcirc 563 - 653$

$653 - 635 \bigcirc 536 - 563$

$-1000 - 76 \bigcirc 76 - 1000$

$-1000 + 173 \bigcirc 827 - 1000$

1. A rectangular tank is 20 cm long and 15 cm wide. It is filled with water to a depth of 20 cm. When a stone of volume 600 cm^3 is placed in the tank, the water level rises. What is the height of the new water level?

2. A rectangular tank is 25 cm long and 20 cm wide. It is filled with water to a depth of 18 cm. When a stone is placed in the tank, the water level rises to 22 cm. What is the volume of the stone?

3. Three identical cubes of edge 20 cm are placed in an empty rectangular tank. 136 liters of water are then used to fill the tank. The width and the height of the tank are 80 cm. What is the length of the tank?

4. Two identical containers A and B have the same amount of water and their water levels are the same. Container A contains 4 marbles and Container B contains 3 metal cubes. When a marble is transferred from Container A to Container B the total volume of water, marble, and cubes in Container B is 300 cm^3 more than the total volume of water and marbles in Container A. What is the volume of each cube?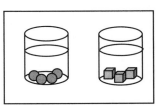

5. A rectangular tank is 20 cm long and 20 cm wide. It is filled with water until it is $\frac{1}{2}$ full. When a metal ball of volume 5600 cm^3 is placed in the container, it becomes $\frac{5}{6}$ full. What is the height of the container?

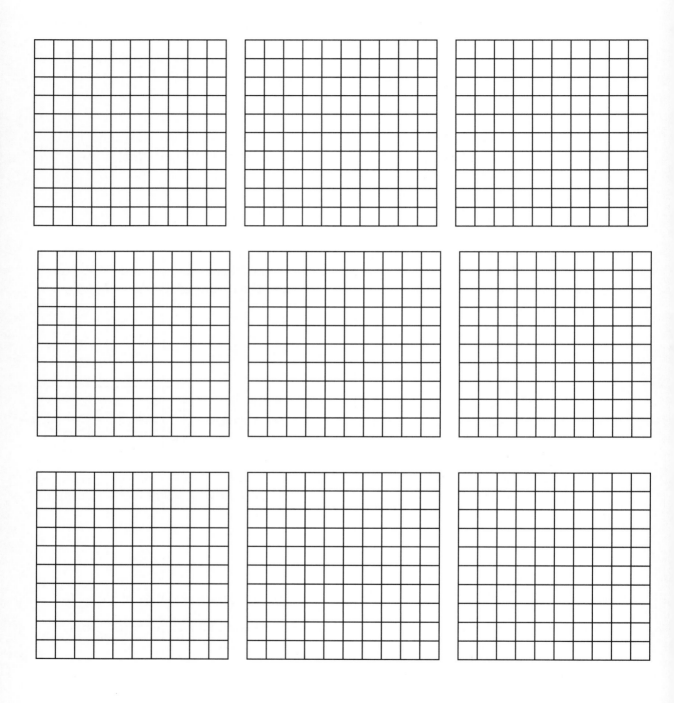

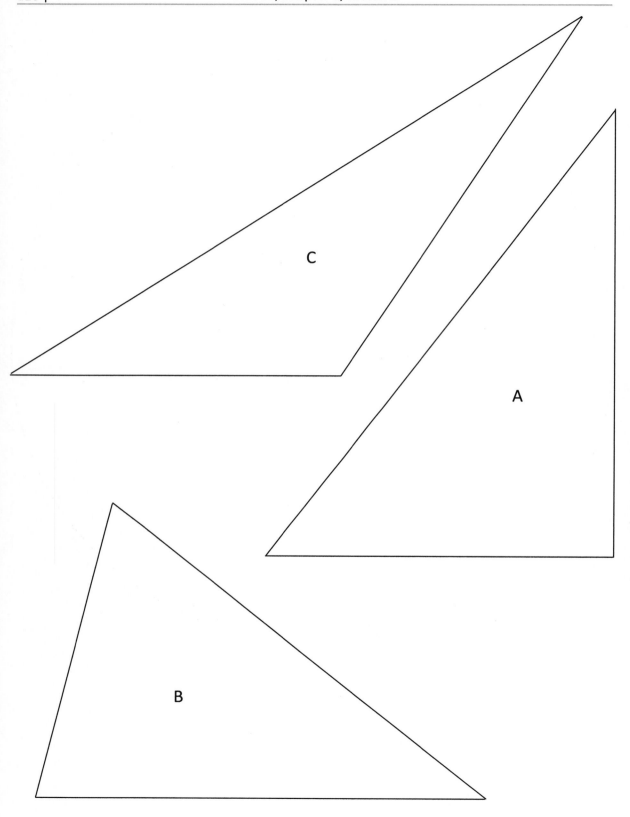

All lines that look like straight lines are straight lines. Find the unknown angles a, b, and c.

1.

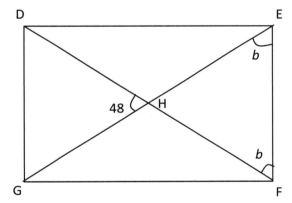

2.

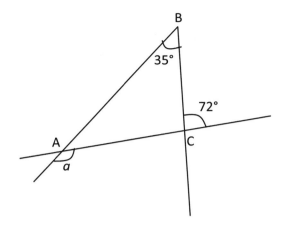

3.

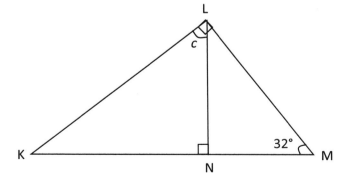

1. Draw a triangle ABC in which AB = 7 cm, ∠CAB = 50°, and ∠ABC = 45°.
    What is the length of side AC?

2. Draw a triangle XYZ in which XY = 6 cm, YZ = 9 cm, and ∠XYZ = 120°.
    What is the measure of ∠XZY? What is the length of side XZ?

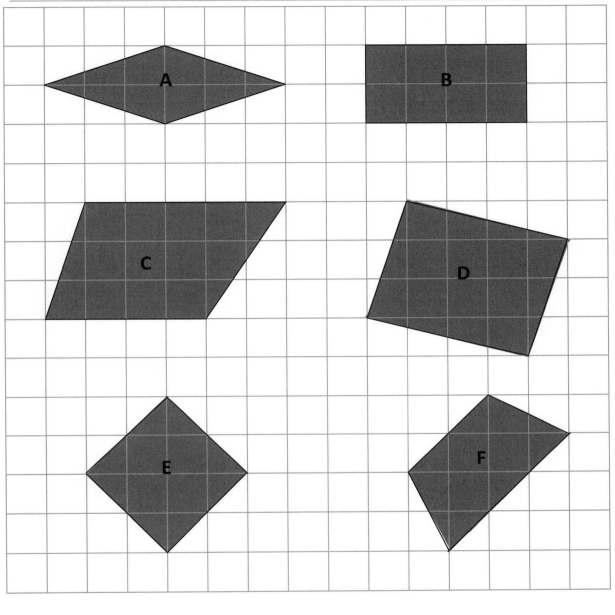

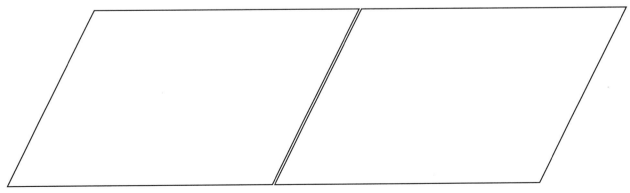

1. A machine makes 1800 cans of soda pop in 1 day (24 hours). How many cans does it make in 1 hour?

2. Water flows from a tap at the rate of 2 liters per minute. How long will it take to fill a 12 liter tank?

3. A photocopier can print 24 pages per minute. How long does it take to print 2064 pages?

1. A machine can make 480 toys in an hour. How many toys can it make in 15 minutes?

2. It takes 10 hours for a carpenter to make 5 chairs. How long will it take him to make 182 chairs?

3. A machine can produce 65 cans of soda in 4 minutes. How long will it take to produce 1300 cans?

4. A wheel makes one third of a revolution in one second. How long does it take to make 1000 revolutions?

5. Sara saves 35¢ a day. How many days will she take to save at least $10?

6. A photocopier can print 16 pages a minute. How long will it take to print 9 copies of a document that is 256 pages long?

1. Water flows into an empty tank at the rate of 8 liters per minute. The water leaks out at the rate of 250 ml per minute. How much water is there in the tank after 2 hours?

2. An empty tank has a base 220 cm by 160 cm and a height of 100 cm. The tank is filled with water from two taps each at a rate of 16 liters per minute. How long will it take to fill the tank completely?

3. A tank measures 90 cm by 50 cm by 70 cm. Five cubes of edge 20 cm are placed in the container. The tank is filled to the brim with water. The water is then drained out from a tap at a rate of 25 liters per minute. How long will it take to empty the tank?

4. Three identical cubes of edge 20 cm are placed in an empty tank. The width and the height of the tank are 80 cm. The tank is then filled with water from a tap at a rate of 8 liters per minute. It takes 77 minutes to fill up the tank. What is the length of the tank?

1. Water flows into an empty tank at the rate of 8 liters per minute. The water leaks out at the rate of 250 ml per minute. How much water is there in the tank after 2 hours?

> Water in: 8 ℓ x 120 = 960 ℓ
> Water out: 250 ml x 120 = 30,000 ml = 30 ℓ
> 960 ℓ – 30 ℓ = 930 ℓ
> There are 930 liters of water in the tank after 2 hours.

2. An empty tank has a base 220 cm by 160 cm and a height of 100 cm. The tank is filled with water from two taps each at a rate of 16 liters per minute. How long will it take to fill the tank completely?

> Volume: 220 cm x 160 cm x 100 cm = 3,520,000 cm^3 = 3520 ℓ
> 32 ℓ $\longrightarrow$ 1 min
> 3520 ℓ $\longrightarrow \dfrac{1}{32}$ x 3520 = 110 min or 1 h 50 min
> It takes 1 h 50 min to fill the tank.

3. A tank measures 90 cm by 50 cm by 70 cm. Five cubes of edge 20 cm are placed in the container. The tank is filled to the brim with water. The water is then drained out from a tap at a rate of 25 liters per minute. How long will it take to empty the tank?

> Volume of tank: 90 cm x 50 cm x 70 cm = 315,000 cm^3 = 315 ℓ
> Volume of cubes: 5 x 20 cm x 20 cm x 20 cm = 40,000 cm^3 = 40 ℓ
> Volume of water in tank: 315 ℓ – 40 ℓ = 275 ℓ
> 25 ℓ $\longrightarrow$ 1 min
> 275 ℓ $\longrightarrow \dfrac{1}{25}$ x 275 = 11 min
> It takes 11 minutes to empty the tank.

4. Three identical cubes of edge 20 cm are placed in an empty tank. The width and the height of the tank are 80 cm. The tank is then filled with water from a tap at a rate of 8 liters per minute. It takes 77 minutes to fill up the tank. What is the length of the tank?

> Volume of water added: 77 x 8 ℓ = 616 ℓ = 616,000 cm^3
> Volume of 3 cubes: 3 x 20 cm x 20 cm x 20 cm = 24,000 cm^3
> Total volume: 616,000 cm^3 + 24,000 cm^3 = 640,000 cm^3
> Length of tank: $\dfrac{640,000 \text{ cm}^3}{80 \text{ cm} \times 80 \text{ cm}}$ = 100 cm
> The tank is 100 cm long.

Sources of Electricity

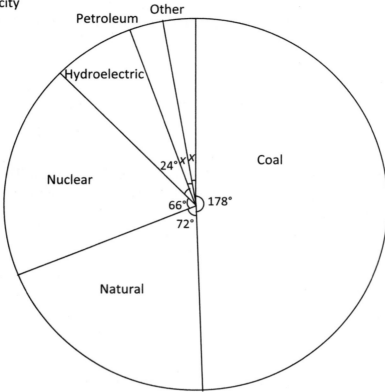

(a) What is the most common source of electricity?

(b) Natural gas makes up what fraction of the sources of electricity?

(c) Petroleum makes up what fraction of the sources of electricity?

(d) Nuclear energy makes up what percentage of the sources of electricity?

(e) If this country uses 1530 billion kilowatt hours of electricity a year, how many kilowatt hours came from hydroelectric power?

Blank Page